W9-BVZ-943

BACK TO THE MOON

BACK
TO THE
MOON

THE NEXT GIANT LEAP
FOR HUMANKIND

JOSEPH SILK

PRINCETON UNIVERSITY PRESS

PRINCETON AND OXFORD

Published by Princeton University Press
41 William Street, Princeton, New Jersey 08540
99 Banbury Road, Oxford OX2 6JX

press.princeton.edu

All Rights Reserved

Library of Congress Cataloging-in-Publication Data

Names: Silk, Joseph, 1942– author.
Title: Back to the moon : the next giant leap for humankind / Joseph
   Silk.
Description: Princeton, New Jersey : Princeton University Press, [2022] |
   Includes bibliographical references and index.
Identifiers: LCCN 2022002240 (print) | LCCN 2022002241 (ebook) |
   ISBN 9780691215235 (hardback) | ISBN 9780691242880 (ebook)
Subjects: LCSH: Lunar bases. | Space colonies. | Space industrialization. |
   Moon—Exploration.
Classification: LCC QB582.5 .S55 2022 (print) | LCC QB582.5 (ebook) |
   DDC 629.45/4–dc23/eng20220420
LC record available at https://lccn.loc.gov/2022002240
LC ebook record available at https://lccn.loc.gov/2022002241

British Library Cataloging-in-Publication Data is available

Editorial: Ingrid Gnerlich and Whitney Rauenhorst
Production Editorial: Kathleen Cioffi
Text and Jacket Design: Karl Spurzem
Production: Danielle Amatucci
Publicity: Sara Henning-Stout and Kate Farquhar-Thomson

Jacket image: The far side of the Moon photographed by Apollo 16,
April 1972. Courtesy of NASA.

This book has been composed in Arno with Neuzeit Grotesk

Printed on acid-free paper. ∞

Printed in the United States of America

10 9 8 7 6 5 4 3 2 1

CONTENTS

PREFACE

WHERE DID WE COME FROM? Are we alone in this vast Universe? I argue that these are questions that only a lunar platform can realistically confront. We need a new science vision for humanity to complement lunar space exploration and exploitation. Such an endeavor can infinitely enrich the anticipated commercially linked activities of the next decades. The outcome can be to attain unique and compelling science goals, if we provide the inspiration. That is the goal of this book. Science-led projects will span our ultimate visions in the world's most ancient sciences, astronomy and cosmology, pioneered in Mesopotamia some 5,000 years ago.

The last frontier in astronomy is exploring the dark ages, before the faint glimmerings of the first light in the Universe. Pristine clouds of hydrogen are building blocks of the future as well as direct witnesses of the past. Very low-frequency radio astronomy is uniquely able to hear their testimony about the beginning of creation, but only if we are brave enough to install low-frequency radio telescopes on the far side of the Moon, the most radio-quiet environment in the inner solar system. Low-frequency radio antennae strung out over large swathes of the lunar surface would tune into the darkest period of our cosmic history, allowing us to unlock the hidden secrets of the beginning of the Universe.

Our complementary goal is the astronomer's holy grail: the search for signs of extraterrestrial life by studying planets around nearby stars. Only a megatelescope on the Moon could reach out to the thousands or even millions of potentially habitable exoplanets we will need to observe in order to have a reasonable chance of detecting the faintest glimmerings of the most primitive forms of extraterrestrial life, and eventually of seeking traces of highly advanced life forms that exceed our own limited capacities. The potential benefits of using lunar resources, of facilitating the construction of scientific infrastructure on the Moon, cannot be underestimated.

Our space telescopes were greatly facilitated by the NASA crewed space flight program, including development of the space shuttle and construction of the International Space Station. Similar logic is likely to apply to lunar exploitation. Lunar infrastructure could subsidize otherwise fiscally unachievable projects such as searching for signatures of remote life using megatelescopes, which are factors of ten larger than any optical telescope feasible on Earth.

The Moon formed as a consequence of a giant impact some four billion years ago with a body that had the mass of Mars. By prospecting on the Moon, we can generate an even deeper understanding of its origin. We will drill deep into the regolith. We will date lunar rocks and measure their composition over a wide variety of lunar terrain. Only in this way can we piece together and reconstruct lunar history. Understanding the origin of our closest neighbor will elucidate the mystery of how the Earth itself formed. Indeed, there are mysteries to be unveiled that will provide clues about the origin of the solar system itself.

Lunar exploration will be humanity's first serious step into space, and establishing habitable facilities on our nearest neighbor in space will be the gateway to a new age of exploration.

There is no technological barrier to establishing an outpost of life on the Moon; with a mix of human and robotic exploration, we can realize that goal. From that unique vantage point, we can move on to explore the possibility of finding life elsewhere in the Universe and probing its beginning. Here is a vision for the next half-century, an exciting time when the major international space agencies are jostling for a habitable space on the Moon. Let's do science!

BACK TO THE MOON

The Moon Beckons

I believe that this Nation should commit itself to achieving the goal, before this decade is out, of landing a man on the moon and returning him safely to earth.

—JOHN F. KENNEDY, 1961

Returning to the moon is an important step for our space program. Establishing an extended human presence on the moon could vastly reduce the costs of further space exploration, making possible ever more ambitious missions. Lifting heavy spacecraft and fuel out of the Earth's gravity is expensive. Spacecraft assembled and provisioned on the moon could escape its far lower gravity using far less energy, and thus, far less cost. Also, the moon is home to abundant resources. Its soil contains raw materials that might be harvested and processed into rocket fuel or breathable air. We can use our time on the moon to develop and test new approaches and technologies and systems that will allow us to function in other, more challenging environments. The moon is a logical step toward further progress and achievement.

—GEORGE W. BUSH, 2004

THE SILVERY WHITE GLOBE in the sky has always fascinated humanity. The Moon is a driver of the ocean tides, but less materialistically and more romantically, the Moon is also a source of inspiration, wonder, even madness. The full Moon glows in reflected sunshine, and the new Moon glints in reflected earthshine. The Moon has no atmosphere. It is lifeless. The lunar surface has craters, mountains, highlands, lowlands, and relics of impacts and volcanic flows. The meteorites and asteroids that have bombarded the Moon over billions of years have reshaped its surface, leaving its features highly weathered by its history of impacts and volcanism. Human traces are few and far between, restricted to a handful of landing sites. But this is destined to change over the next decades as human lunar exploration and exploitation projects get underway.

As the closest heavenly body to us, the Moon is a special place and has been the source of speculation since time immemorial about the possibility of extraterrestrial life. Thanks to decades of lunar surveys, we now know that it is not the most welcoming of environments, but humanity has acquired the capability to modify its environment. There are serious prospects for lunar habitations over the next few decades.

A space race has begun anew, half a century after Neil Armstrong's first giant step for humankind. To prepare the pathway for the future of crewed spaceflight to the Moon, numerous lunar landings have been made. NASA currently plans to build a lunar orbiting space station, the Lunar Gateway. This station will coordinate lunar development and be a launch site for solar system exploration. The international space agencies plan to build moon villages for habitation. There are plans to tap commercial interests to develop luxury hotel resorts. As terrestrial resources of the rare earth elements and semiconductor materials become sparse, lunar mining operations are also envisioned. And ultimately lunar rocket fuel will launch us further into the solar system.

The habitable facilities developed for lunar exploration will enable exploitation of lunar resources and help to demonstrate the feasibility of the next great chapter in interplanetary exploration, which will take us to Mars and beyond.

Lunar geology will guide us toward an understanding of how the early Earth itself evolved if we can first confirm the origins of the Moon. The leading theory invokes a major impact with a body the size of Mars some 4.5 billion years ago. The backsplash debris from this collision condensed to form the Moon. The initial collision most likely misaligned the Earth's rotation axis, and it is this misalignment by 23.5 degrees that gives us the seasons and enables the Earth to avoid extreme climate variations. Confirming the origins of the Moon through lunar prospecting will inevitably lead us to a better understanding of the origin of the Earth, and indeed of the origin of the solar system itself.

For any explorer who dreams of leaving the Earth, the Moon is the first destination. A mere quarter of a million miles away, it takes just three days to get there. And at least for the moment, the Moon is pristine territory.

Half a century has elapsed since man last set foot on the Moon. Human exploration stopped abruptly after the era of the giant Saturn V rockets. Launch resources are gradually being restored, through both commercial and international pressure, and several countries are developing ultraheavy launchers. Within a few years, the scene will be set for a new phase of human lunar exploration.

Lunar development was always subject to the maturity of the program of space travel and exploration, and especially to the budget. Today the political mood is changing, and the Moon is now being considered by the international space agencies for a variety of uses. Most are commercially driven, but some are purely scientific. This is good news for science, as there should be new opportunities to explore the Universe by riding on the coattails of planned lunar developments.

The unlikely combination of events that formed the Moon suggests that Earth-like life is a rare phenomenon in the Universe. There is no way to evaluate the odds of such special configurations, however, until we begin to search for them. And the Moon itself may provide the ideal platform for initiating our search of the nearby Universe.

A baby's tentative first step marks a transition in human evolution. The analogous step in biological evolution was the emergence of *Homo sapiens*, which in its final form took something like 200,000 years to be accomplished on Earth. This is such a brief flash of cosmic time in the past billions of years that we can anticipate that the process may have been duplicated elsewhere in our galaxy. The odds are incalculable, however, and the biology simply too complex. Our only resort is to search for the process elsewhere.

The Moon will enable us to forge new paths in space exploration by providing a combination of infrastructure and telescopes of unprecedented power. Inevitably, it will furnish a site for launching solar system probes. Spaceports on the Moon and in orbit around it are essential for launching heavy loads. We cannot realistically achieve crewed solar system exploration from the Earth: the fuel requirements to escape Earth's gravity are too costly. The Moon will provide the practically unlimited local fuel supply necessary for space travel to explore the solar system. Setting up a permanent lunar base will be humanity's first cautious step toward the stars.

Eventually, the stage will be set for interstellar exploration. Whether through cryogenic conservation of humans or genetic rebirth, the century-long travel times to the nearest stars will not deter future generations of astronauts. Indeed, once computers greatly outperform human brains, the interface between robots and humans will seem fairly fuzzy. The future of computing power currently seems unlimited as new innovations are developed. It would be rash to draw any lines in the sand that might limit the future that beckons—our potential for exploration and population of the cosmos.

The low gravity on the lunar surface will aid the construction of facilities for habitation. Already, buildings are being designed to be situated near the lunar south pole, where the shadowed depths of numerous deep craters are in permanent darkness. Polar craters are shielded by the low elevation of the Sun, which does not rise above the crater rims. The range of temperatures is more stable at the poles than on the lunar equator. The latter varies from minus 180 degrees Centigrade to 130 degrees Centigrade. The permanently shadowed craters contain large deposits of ice.

On the crater floor, the permanent shadowing allows stable temperatures of around minus 200 degrees Centigrade. The coldest temperature ever recorded on Earth is about minus 90 degrees Centigrade, recorded at ground level at Vostok in Antarctica. The highest temperature recorded, in Death Valley, is about 70 degrees Centigrade. These conditions would be tolerable for robotic activity, but human presence would require sheltering. Power sources are readily available for local cooling or heating. The crater rims are almost perpetually illuminated by sunlight and provide an inexhaustible source of solar power during the long lunar nights.

A next step would be to develop habitats situated in vast lava tubes that would be accessible from the lunar surface. These caverns could be large enough to host and naturally shield entire cities from any life-threatening solar-induced activity. Cities could be constructed in giant lava tubes to provide shelter from meteorite impacts, and protection would also be provided against occasional violent solar flares and bombardment by micrometeorites.

Lunar mining may provide a limitless supply of rare earth elements, and robotically aided extraction and ejection of toxic debris into space would limit the pollution associated with this activity.

Local production of rocket fuel will make lunar and cislunar environments optimal for developing future launch sites for interplanetary space probes. Extraction of rocket fuel is planned in the form of liquid hydrogen and oxygen. The fuel would be sourced from ice deposits in cold polar craters.

To take advantage of these many planned activities we need to promote science as a key goal of lunar exploration. We can learn about the origin of the Moon and the origin of the solar system in this unique environment, whose low gravity and lack of an atmosphere will enable projects on scales that are unimaginable on the Earth.

Conditions for living on the Earth are rapidly deteriorating. Global warming, pandemics, environmental degradation, depletion of natural resources, overpopulation, and wars are taking their toll. It is only natural to look elsewhere. In the past, frontiers beyond Europe expanded to Africa, to the Americas, and beyond, at immense cost to the indigenous populations. Sooner or later, we will need to expand our frontiers beyond the Earth. The lunar surface is not the ideal location for a high-density human population, but the Moon does provide a base for humanity to prepare for a future of unprecedented exploration as well as for the consequences of potentially devastating catastrophes.

The commercial backdrop opens a highly subsidized window for doing science on the Moon. We can make a compelling case for leveraging telescope construction, which is relatively cheap and opens up new vistas of human exploration. Robots and humans can work together to build lunar telescopes of unprecedented size and sensitivity beyond those of any terrestrial or space platforms. Lacking an atmosphere or ionosphere, the lunar surface can provide a stable platform for new generations of giant telescopes that will peer deeply into the remotest depths of the Universe and also relay images of nearby planetary systems.

One of the most fundamental questions we ask about the Universe is whether we are alone. If life is common in the Universe, why haven't we encountered aliens? There are billions of Earth-like planets in our galaxy, and we know that around 50 percent of them are much older than the Earth because their host suns are billions of years older than our Sun. But so far our telescopes are far too small to do more than scrape the surface of this potential treasure trove of information on other planets. We know what to look for: the reflected glints of oceans, the green glows of forests, the presence of oxygen in the atmospheres, and even more advanced but subtle signals of intelligent life such as (hopefully transient) industrial pollution of planetary atmospheres.

However, even the existence of alien life is debatable. We have absolutely no idea of the odds of encountering advanced life forms on the billions of habitable planets in our galaxy. To move from science fiction to facts, our only viable strategy is to look, and lunar telescopes will enable us to take a first tiny step in exploring our galaxy for signatures of life.

Lunar megatelescopes can be sited near the poles of the Moon in deep craters that provide total darkness and a spacelike chill that is ideal for infrared astronomy. These sites have no atmosphere to absorb the infrared light from space, and there is a permanent nearby source of solar power. In low gravity, structures can be built that are vastly larger than their terrestrial counterparts.

Certain questions have obsessed humanity for eons. The first conscious beings looked at the night sky and marveled at the glory of the Moon and the limitless saga of countless twinkling points of light. Where did we come from? This is the question posed by ancient philosophers from the dawn of civilization. To make progress, we must target an epoch we know next to nothing about—the dark ages of the Universe, before the first stars form, when there is no light but trillions of primitive hydrogen gas clouds.

The first clouds are primitive fossils, teeming with primordial radio whirls. If our telescopes can gain enough sensitivity to detect them as infinitesimal shadows against the leftover radiation from the Big Bang, we will be able to probe the first instants of the universe. The lunar platform is the last stop for astronomy. It is the ultimate site for building low-frequency telescopes of unprecedented aperture and power to accomplish these challenging goals.

After the epoch of the first clouds, more complex structures rapidly developed. Evolution was inexorable, largely thanks to the attractive power of gravity. Soon, after only a few million years had elapsed, the first galaxies and the first massive black holes were in place and there was no looking back. Telescopes on the Moon will probe these epochs more powerfully than any conceivable terrestrial or even orbiting space telescope.

Scientific ventures are likely to accompany lunar exploration. Of course the first priority is to lay out the infrastructure for the long term, so that we can take a first step toward space colonization after establishing permanent lunar colonies. Massive payloads, which are energetically unfeasible for launch from terrestrial sites, can be launched from the Moon or from cislunar space. The lunar platform provides our only sustainable pathway into space exploration. As discussed later, NASA already envisages building an orbiting lunar space station and spaceport. From the Moon, we can reach out to the solar system and ultimately to the stars. Our long-term goal must be to venture into space to develop new horizons for humanity.

Lunar telescopes will open new horizons in our quest to understand the Universe. For a small fraction of the cost of a crewed return to the Moon, we could build lunar telescopes of unprecedented size and observation clarity compared to our limited terrestrial vision. We can answer questions of fundamental importance about our origins. We can look for indications of extraterrestrial life.

Certainly, commercial ventures will have priority in lunar exploitation and be a major driver of lunar exploration. We will inevitably need to balance the demand for tourism with industrial activities. Human and robotic activities will be closely linked as well. As we look ahead, establishing lunar settlements is inevitable if we are to do science on the Moon. This will be one of our first lunar activities to enable science to expand the boundaries of what we can and cannot do.

Outer-space legal treaties will need to be signed by all countries undertaking lunar surface exploration and exploitation, including those already planning such activities: the United States, China, Russia, and Europe, as well as India and Japan. The worst-case scenario would be a lunar nightmare with free-for-all lunar exploitation reminiscent of the Wild West. We urgently need legal treaties to cover everything from real estate to crime, mining rights, and military and scientific activities. We will need to deploy multinational space forces to enforce laws, and lunar mines, spaceports, and even telescopes will also require international regulation.

We will go back to the Moon. We will build on the Moon. We will colonize the Moon. We will exploit the Moon. We will do science on the Moon. Lunar science will open new vistas on the most profound questions we have ever posed. And above all, humanity's drive for exploration will set the final frontiers. This book describes that journey into the future.

CHAPTER 1

The New Space Race

[Space travel] will free man from his remaining chains, the chains of gravity which still tie him to this planet. It will open to him the gates of heaven.

—WERNHER VON BRAUN

The ultimate purpose of space travel is to bring to humanity, not only scientific discoveries and an occasional spectacular show on television, but a real expansion of our spirit.

—FREEMAN DYSON

The Past Half-Century

In 1961 President John F. Kennedy decided that the National Aeronautics and Space Administration (NASA) would land men on the Moon by the end of the decade. His decision was in part a reaction to the first space flight around the Earth by the Soviet cosmonaut Yuri Gagarin. The United States needed to counter. The Soviets already had made a head start in planning for a space station and orbital flights in space. A crewed lunar landing was the most feasible step that the United States could take to reinforce its superiority in space. And national security issues were certainly a factor in the president's decision.

Kennedy took up the gauntlet in an address to the nation a year later, on September 12, 1962:

> We choose to go to the Moon in this decade and do the other things, not because they are easy, but because they are hard; because that goal will serve to organize and measure the best of our energies and skills, because that challenge is one that we are willing to accept, one we are unwilling to postpone, and one we intend to win.

And so began the Apollo program, approved by the US Congress in 1963 to send American astronauts to the Moon. Kennedy, sadly, did not live to see his promise to the nation fulfilled.

Various US presidents had weighed in on space exploration before Kennedy. It was Dwight Eisenhower who initially responded to the surprise Soviet launch of Sputnik in 1957. He decided that US space activities should be administered by a civilian agency. Though he assured the public that Sputnik was but "one small ball in the air," Eisenhower nevertheless saw to it that NASA was established the following year. Spacecraft design and launch facilities were organized. The Mercury program was established to evaluate the feasibility of low near-Earth crewed orbits as a precursor to going further into space. The program culminated in the first crewed suborbital flights.

When Eisenhower set up NASA in 1958, the space race was on. Crewed low-Earth orbital flights would set the pace for human space travel. In 1961, Yuri Gagarin was the first man in space when his Vostok spacecraft took about 108 minutes to orbit the Earth once before landing by parachute in the Soviet Union. His flight was followed within a month by the first suborbital spaceflight by an American, Alan Shepard, in a Mercury capsule that flew for just 15 minutes.

In the words of the spacecraft pioneer Wernher von Braun: "To keep up, the USA must run like hell."

Mercury-boosted orbital flights soon followed. The Mercury missions led to Earth orbital crewed flights. The first flight was three Earth orbits in 1962 by John Glenn, who would go on to serve for twenty-five years as a US senator for Ohio. The Mercury program culminated in 1963 with a daylong flight by astronaut Gordon Cooper. It was succeeded in 1965 and 1966 by the Gemini orbiter, which carried two astronauts in low-Earth orbit for prolonged space missions. Humanity was preparing to go beyond the Earth.

The United States was determined to catch up with and overtake the Soviet Union, and the Moon clearly was the next goal. Eisenhower was reluctant, however, to enter into a space competition with the Soviet Union. The space race only really began when President Kennedy laid down the challenge in 1961. His goal was clear: human landings on the Moon. The Moon had remained a distant dream for crewed missions until Kennedy's intervention, which led to the establishment of the Apollo program.

Of course, robotic space missions to the Moon came first. The Soviet Union flew the first sequence of robotic lunar landers, the Luna series, which ran from 1959 to 1976. In all, there were seven Luna soft landings. The Soviet space program did not culminate in crewed landings in large part because of the Soviets' failure to develop a heavy-load spacecraft in time to compete with NASA. The Soviet space agency had begun with a strong lead in low-Earth orbital space flights but could not keep pace with the Apollo landings between 1969 and 1972. As the technical gap between the two countries grew, political considerations in the Soviet Union became more important. Its crewed lunar program was closed down in 1974.

The US crewed lunar missions were carried out in several stages. First, a series of robotic missions were conducted to prepare the way. NASA's lunar Ranger program began in 1961 with a spacecraft equipped with television cameras to record possible landing sites. After a series of crash landings on the Moon, it was superseded five years later by the lunar Surveyor program. These spacecraft soft-landed on the Moon and studied its surface composition. A new series of lunar orbiters was launched, beginning in the same year, to search for possible landing sites for the upcoming crewed missions.[1]

The Apollo program achieved success in 1969 with the first crewed lunar landings. All Apollo landings were on Saturn V launch vehicles. The three-stage Saturn V spacecraft was 110 meters high, or 15 meters taller than Big Ben, and its total weight was 3,000 tons. More than 90 percent of the weight was liquid propellant fuel. The third stage was fired from low-Earth orbit, launching the Apollo spacecraft to the Moon. It carried a 50-ton deliverable payload, including the lunar lander. Also on board was the Command and Service Module, which remained in lunar orbit to return the astronauts safely to Earth. The highlight of the first phase of human lunar exploration was the 1969 landing of Apollo 11 in the Sea of Tranquility. Neil Armstrong and Buzz Aldrin performed the first moon walks before a worldwide audience.

In all, there were six crewed Apollo missions that landed on the Moon. Each carried three astronauts some 239,000 miles into space and then returned home safely. Although the United States would continue to dominate lunar space exploration for decades, the crewed lunar program was over in less than four years. The last crewed mission to the Moon was in 1972. Saturn V last flew in 1973, when it launched Skylab, a precursor to the International Space Station.

Skylab's orbit slowly decayed and the space station disintegrated in 1979, on reentry to the Earth's atmosphere. The shower of debris covered western Australia and parts of the Indian Ocean. Its successor, the International Space Station, would not be launched into low-Earth orbit until 1998. The ISS features a still ongoing international collaboration. Russia launched the first module of the ISS on a Proton rocket from the Baikonur site in Kazakhstan, and months later the first crewed flight to ISS was made by the Endeavour orbiter, the fifth and last spacecraft of NASA's Space Shuttle Program. Long-duration visits began in 2000.

The Apollo program culminated with six successful lunar astronaut landings, but support for the program did not continue. Lyndon Johnson's Great Society Program trumped space spending, at least for any human lunar exploration, and the huge space budget could not be justified. With the end of the era of heavy lifters, Saturn V proved a hard and expensive act to follow. Nevertheless, only twelve years after Sputnik's low-Earth orbit, some 300 miles above the Earth, the United States had won the race to the Moon.

Return to the Moon

Today the major space agencies are gearing up to return to the Moon. They will spend decades surveying the surface and then move on to the deployment of resources. Will the enormous costs involved inhibit or limit the lunar exploration program? If there are cutbacks in government funding, are science projects doomed? Will military prerogatives hold sway? Commercial goals are likely to dominate, with lunar science inevitably suffering, but there is hope. Space is limited on the lunar surface, at least of the quality needed for major telescope projects, but the space available should suffice.

Plans to return to the Moon are now becoming serious thanks in large part to international competition for lunar resources.[2] An optimistic outlook on this competition is that science and activities such as tourist travel and mining will share lunar resources, and that science will benefit from the necessary infrastructure. For this complementarity to prevail, however, and for the pristine lunar environment to be preserved, international agreements will be needed.

Successive US presidents touted the value of space exploration but settled for robotic ventures that explored the solar system. Cost was an overriding issue. The situation was to change only when serious international competition came to the fore. Now, with that competition coming from China, there once more are military considerations, but they remain on a backburner, as commercial aspects loom large. The immense interest in space tourism and lunar mining ventures may seem futuristic, but the spoils may go only to the first arrivals.

There was a moment of real international collaboration in the decade following the Apollo years. The space shuttle was developed during the Nixon administration. Later, in 1984, under Ronald Reagan's leadership, planning and construction began on an orbiting space station. It would take more than a decade to complete.

As costs grew uncomfortably high by the early 1990s, the incoming administration of Bill Clinton and Al Gore had to make a difficult financial decision. To limit the drain on the NASA budget, they decided to bring in new partners, and the new orbiting space station became the International Space Station. Soviet and NASA astronauts shared the crewing and the transport costs, and other international agencies that nursed space ambitions joined as partners, including those in Europe, Canada, and Japan. The first joint US-Russian crew began living in the ISS in 2000, and it has been crewed ever since.

But the International Space Station, moving in near-Earth orbit, eventually must be superseded. It has always been an outstanding laboratory for training astronauts in preparation for more distant space exploits. After the Space Shuttle Program was phased out in 2011, NASA contributed to the space station servicing missions by purchasing space on Russian spacecraft. Then new plans were developed and US ambitions became grandiose. President George W. Bush announced in 2004 that NASA's human space flight program would start "with a human return to the Moon by the year 2020, in preparation for human exploration of Mars and other destinations."[3]

With successive US administrations, the time line for human space flight has been prolonged. And as realism sets in, so has the time line for any final destination beyond the Moon. A major change has now come about with the development of commercial spacecraft. Elon Musk's SpaceX Falcon 9 was the first commercial spacecraft to service the International Space Station. Deployment of a cargo spacecraft containing some three tons of supplies and experiments soon followed. The Dragon crew spacecraft transported four astronauts to the ISS in November 2020. Commercial exploitation of human spaceflight has truly begun, and commercial missions to the Moon will surely follow.

At the same time, China has announced plans to build a new space station in near-Earth orbit. This is far from China's only goal. Its National Space Administration is pursuing a human outpost on the Moon, among other lunar projects. Not to be outmaneuvered, President Donald Trump declared that the next time US astronauts blast off, they would be headed to our rocky satellite. In 2021, his successor, Joe Biden, endorsed

FIGURE 1. How light, water, and elevation will be provided at NASA's Artemis base camp on the Moon. American astronauts will take their first steps near the Moon's south pole, a land of perpetual light, extreme chill, darkness, and frozen water. NASA's next leap forward into interplanetary space will be fueled there, perhaps by 2026. Through the use of lunar rovers with 3-D printing capacity, power will be generated from solar cells and food will be produced in greenhouses.

Image credit: Image by P. Carril in European Space Agency, "ESA Opens Oxygen Plant—Making Breathable Air out of Moondust," January 17, 2020, https://scitechdaily.com/esa-opens-oxygen-plant-making-breathable-air-out-of-moondust/.

NASA's Artemis program to operate an orbiting space station along with lunar bases tapping the low gravity of the Moon. Artemis will be permanently crewed and is designed to facilitate space travel throughout the solar system. Biden also gave his support to the newest branch of the US armed forces, the United States Space Force.

Both China and private entrepreneurs in the United States are enthusiastic about mining minerals on the Moon. There are high-value resources to be extracted, including the rare earth elements and semiconductor materials for which terrestrial mining capacity is limited. One key project will be production of rocket fuel out of lunar ice, a prerequisite for further space exploration. Large lunar living habitats are being planned. The European Space Agency has called for the installation of a permanent, inhabited village at the lunar south pole. As a first step, a program has begun of synthesizing lunar construction materials with water and regolith, which is naturally available on the Moon.

What has happened post-Apollo? China is leading the current Moon rush. The first robotic Chinese probe to the Moon landed in 2013. Chang'e 3 carried a rover that operated briefly in Sinus Iridum, an area of dark lava flows in the lunar highlands. This area is being considered for future colonization in the nearby giant lava caves. The first landing on the far side occurred in January 2019. A lunar rover on board Chang'e 4 landed in the huge Von Kármán crater, which is some 186 kilometers across. This is part of the South Pole–Aitken Basin, which measures about 2,500 kilometers. As we will see, lunar polar craters especially have commercial and scientific potential.

Meanwhile, NASA has been scrupulously conserving frozen and vacuum-packed samples of retrieved lunar soil, but some of the Apollo-collected soil samples taken fifty years ago show some deterioration, likely caused by water vapor contamination. NASA has a monopoly on exploration of the lunar surface, but for how much longer? China is rapidly catching up. At the end of 2020, the Chang'e 5 mission conducted a sample return of lunar rocks to Earth. About 2 kilograms of rocks were delivered to a landing site in Inner Mongolia.

The lunar terrain covered in the early years by Apollo 15, 16, and 17 was limited. We need to sample new sites with more varied geological records. Pristine new samples are expected to help focus efforts to better understand the geological context and history of lunar soil and rocks. As we acquire new insights into how the Moon formed, we may incidentally learn about the potential for lunar mining.

The Next Stop for Humanity

There is a clear path forward for lunar exploration and exploitation.[4] It may take up to a century to get there, but the process has begun and all major space agencies are enthusiastically joining the race to the Moon. What should we do once we settle on the exploration program?

There are lunar resources to be mined. The Earth is exhausting its easily accessible supplies of ores such as rare earth metals, which play a crucial role in the electronics industry. There may well be centuries of reserves on Earth, but we should be taking the long-term view. Mines can be toxic environments. Avoiding pollution will be essential. The lunar resources offer other challenges, but environmental protection will play an important role if built in at the onset.

Our visions for future exploration of the Moon already include plans for industrial applications, ranging from manufacture in low-gravity environments to mining of rare earths and fuel production for interplanetary travel. There is a huge demand for lunar projects, in large part because the opportunities for entrepreneurship are unparalleled. These opportunities, pursued through both human and robotic exploration, will be realized by the mid-twenty-first century.

The Moon presents a potentially vast tourist industry. As humanity seeks new challenges, the Moon offers dazzling new horizons for leisure and sports activities. The commercial aspects of these activities will drive investment in them. Mass transportation for human lunar travel will not be offered in the first decades—not until the pent-up demand for luxury tourism is met. The Moon will initially be a playground for the rich, but change over time is certain once low-cost space transport systems are developed. Giant lunar parks for leisure and relaxation will be established, and low-cost housing will be designed to host the necessary support personnel to organize mass tourism. In the next half-century, the Moon seems destined for such activities, with commercial backing.

The Moon's low gravity and the presence of water ice have inspired designs for lunar habitations. The Moon may not be the ideal solution for the Earth's overpopulation problem, as much of the surface is a hostile environment. One cannot imagine that the lunar surface could accommodate large numbers of permanent inhabitants apart from key workers. However, it does offer some enticing pieces of real estate with relatively moderate climates, especially in the polar regions. And inevitably, the wealthiest segments of terrestrial society will find it hard to resist the attraction of the Moon as a new environment for developing secondary residences.

There is lots of lunar dust, accumulated by meteorite collisions over billions of years, and its abrasiveness creates an environment not ideal for the smooth running of machinery. Although the dust could be easily lifted above the surface by entrepreneurial activities, it makes for a potentially dangerous environment, especially with respect to the pulmonary health of long-term lunar residents. It will be essential to develop effective filtering once we build lunar telescopes.

One positive aspect of dust is that it furnishes an ideal material for developing robust building materials. With its unique reservoir of rare elements, lunar dust allows unique commercial mining ventures to become feasible. Unique science projects should accompany them.

Power is a crucial accompaniment to surface development. One advantage of the lunar surface is the abundance of solar power it can provide. Perhaps the first sites for development will be the polar craters, whose high rims are permanently illuminated by sunlight. Polar craters are the preferred sites for developing the first lunar bases because many of them are in permanent shadow—which also makes them attractive sites for telescopes—yet have an inexhaustible supply of nearby solar power on the high crater rims where the sun never sets, and their polar ice deposits enable in situ construction facilities.

Moonquakes are caused by a slight shrinking of the lunar crust over millions of years, along with the buildup of stresses by the action of the Earth's and Sun's competing tidal forces on the stretching of the lunar crust. Perhaps the crust stretches one-tenth of a meter over 100 years, but that is enough to produce wrinkle-like faults on the surface as well as shallow moonquakes. Their strength, which ranges up to 5 on the Richter scale for earthquakes, can be measured by seismometers laid out by the Apollo astronauts.

Major lunar quakes seem to be rare. There are hundreds of weak moonquakes per year—down to 2 on the Richter scale—but there are millions of similar-strength earthquakes every year. The lack of extensive tectonic lunar activity means that the surface is seismically stable. That is good news for the construction of large telescopes.

The Moon has many advantages for astronomy over a free-flying telescope in space. The large, cold, and dark polar craters are preferred sites for infrared astronomy because the lunar atmosphere does not degrade the spectral and seeing capabilities of telescopes. The entire electromagnetic spectrum is available for exploration, from the ultraviolet to the far-infrared bands. The lunar environment, with its low lunar gravity, provides an ideal platform for constructing a new generation of really large telescopes.

The ionosphere is a tenuous layer of ionized gas in the Earth's outer atmosphere. It presents a huge distraction for radio waves, especially at the lowest frequencies, which is where we expect the most interesting signals from the very early Universe. Low-frequency radio waves are deflected and scattered by the ionosphere. The Moon has a negligible atmosphere. The far side of the Moon, shielded from the Earth, is an ideal site for radio telescopes and opens up a new window on the Universe because it allows us to do very low-frequency radio astronomy that is unfeasible from the Earth.

Let's go back to the Moon! Our space agencies are intent on developing lunar bases to support activities ranging from mining to spaceports, but building telescopes to explore the beginning of the Universe could be done at a modest fraction of the lunar infrastructure cost. Not only could we view the cosmos in a way that is currently unimaginable from the Earth, or even from space, but lunar platforms also have the potential to be immensely rewarding for fundamental physics goals. However, the planning must begin soon.

The race to the Moon is opening up as different space agencies join the rush. The first evidence for water on the Moon was discovered by a US-built experiment on the Indian Space Research Organization's Chandrayaan-1 spacecraft, launched in 2008. Planetary geologist Carle Pieters of Brown University led a team that detected infrared signatures of water molecules in the sunlit lunar soil. They mapped hydroxyl molecules—building blocks for water—that most likely had dispersed from layers of ice in cold, dark craters near the poles.

This lunar orbiter gave a huge boost to India's autonomous space program. ISRO scientists led by Ashutosh Arya also carried out 3-D surface imaging and discovered an empty giant lava tube (or cave), two kilometers long and some 160 meters below the lunar surface. The opening of the cave, which is 120 meters high and 360 meters wide, formed from the uncollapsed remains of an ancient volcanic lava flow. The 40-meter-thick roof of the lava cave offers protection from ultraviolet radiation, dust, micrometeorites, and extreme temperature variations, at a temperature of minus 20 degrees Centigrade. Lava tubes are naturally protected environments that are potential sites for future lunar bases.

In the United States, lunar space activity saw a revival in 2009, when a reconnaissance lunar orbiter was launched by NASA on the two-stage, 600-ton, medium-lift vehicle Atlas V. The impact of the accompanying module lifted plumes of crater debris miles above the surface. These plumes were found to be rich in volatile ices, including water, thus confirming the evidence for water that was earlier reported by the Chandrayaan space mission. More recently, in 2021, the SOFIA airborne far-infrared telescope found evidence of water molecules most likely dispersed throughout the regolith. In fact, these water molecules were detected in one of the largest visible craters and amounted in concentration to a bottle of water trapped in a cubic meter of soil spread over the surface. The fraction of water observed was only a percentage of the water detectable in the Sahara Desert, but there is far more desert on the Moon.

There had already been hints of lunar ice a decade earlier. One of the instruments on board the NASA Lunar Prospector spacecraft, a neutron monitor, scanned many craters around the lunar south pole. The idea was to look for neutrons produced by cosmic ray bombardment of the surface. The detected deficiency of neutrons is considered to be a proxy for hydrogen atoms—which of course do not contain neutrons—and hence water. It seems that there is abundant water ice on the Moon in the polar craters, which act as cold traps for impinging material from cislunar space.

The Lunar Gateway

Humans will certainly return to the Moon in great numbers, and an ultra-heavyweight launcher will be crucial for crewed landings. The huge cost of Saturn V ($50 billion in 2021 dollars) explains why half a century has passed before we are on the verge of flying comparable lunar payloads. The United States and China are currently leading the way, and Russia and India are also planning crewed missions to the Moon. Of course, the United States has other priorities much closer to home. The Apollo space program amounted to about 4 percent of the US annual federal budget at the time. One cannot imagine a comparable effort today.

Future lunar science, exploration, and utilization will build on current and upcoming automatic and planetary robotic missions. A flotilla of lunar orbiters has been deployed by several countries for science and reconnaissance in the past decade. The orbiting robotic spacecraft are providing new views of the Moon, as well as its environment and resources. Several international space agencies are pursuing lunar mapping programs. The main aim is to prepare for crewed missions a decade from now whose goals will include reconnaissance of future landing sites. Prospecting for mineral resources is the likely endgame.

Competition is energizing future plans, and transport to the Moon is a crucial factor in carrying them out. NASA has announced plans for commercial development of crewed launch vehicles that will transport construction materials to the Moon, and it is actively seeking private companies to build future launch vehicles and lunar landers. Three companies were selected in 2020 to compete for contracts worth $1 billion to develop the hardware for crewed landers, and Elon Musk's SpaceX was retained in 2021. Developers of commercial transport, which could prove profitable over the long term, are raring to go to the Moon.

The immediate future in crewed space missions to the Moon lies with NASA's Space Launch System. SLS is intended to be the first launch vehicle in half a century to have a liftoff weight comparable to that of Saturn V. The first crewed flight of the new heavy launch vehicle should occur in 2023. Astronauts will arrive at the lunar outpost in a crew capsule being built by Lockheed Martin. The European Space Research Organization will deliver the service module, all to be piggybacked onto the launch spacecraft.

Adequate launch capability for ultraheavy loads is a prerequisite for establishing the infrastructure for lunar base construction. Many launches will be needed to put the essential infrastructure in place so that human activities on the Moon can be developed over the next decades.

An early step will be to install the Lunar Gateway, the orbiting lunar space station that is a key component of NASA's Artemis program. Starting in the mid-2020s, the Lunar Gateway will support a succession of crewed lunar launches and landers. The space station will initially have a minimal number of modules, sufficient to send shuttles to the surface, ensure a safe lunar landing, and return.

A long-term goal of the Lunar Gateway will be to send crewed probes to Mars. In view of the technical difficulties for human travel that are anticipated for prolonged space trips (described later), such crewed probes are certainly decades away. In the meantime, the Lunar Gateway will focus on lunar exploration.

Within the following few years, NASA expects to develop an enlarged lunar space station capable of deploying many astronauts to the lunar surface. The Lunar Gateway is intended to be a space traffic hub with multiple docking ports to launch crewed lunar landers. Astronauts will work on the surface and return to the lunar space station as their residential base. A permanent and sustainable human presence in lunar orbit is expected by 2028.

The orbiting station will also serve as a spacecraft refueling station to facilitate travel back to Earth and beyond. Crewed missions to the lunar surface will build outposts, laboratories, and surface observatories in order to develop the extensive infrastructure needed for local construction, transport, and deployment of activities on the lunar surface.

The International Competition

NASA is not the only player in the race to land humans on the Moon. Various national space agencies are vigorously carrying out plans to go to the Moon, most notably the Chinese space agency, the China National Space Administration (CNSA). The first post-Apollo astronauts are likely to land on the surface near the lunar south pole in 2024. NASA has announced that the lunar astronauts will include a woman and also a person of color. Meanwhile, the European Space Agency has announced its goal of further diversifying the future astronaut pool with disabled candidates.

NASA clearly has competition. Since the Apollo era, the United States has been relatively cautious in reinvigorating the planning for crewed lunar missions, and that has provided a window of opportunity for its rivals—most notably China—to catch up. China is the third country to have sent humans into near-Earth space, beginning with Yang Liwei's successful twenty-one-hour flight in 2003 on the Shenzhou 5 launcher. No doubt an element of healthy competition will arise between China and other countries now that it has announced its intention to place the first woman on the Moon in the 2020s.

An ambitious lunar exploration program is being undertaken by the Chinese space agency, which is currently designing an ultraheavy-load spacecraft to go to the Moon. The first flight of Long March 9, with a 100-ton payload capacity, is scheduled for 2030. A crewed mission is planned in 2036, to build an outpost near the lunar south pole.

Russia is intent on overcoming any lingering inferiority complex from the Apollo decade by fully engaging in the new space race to the Moon. Moscow had several breakthroughs to its credit in the early days of the space race, including the launch in 1957 of the first artificial earth satellite and Yuri Gagarin making the first journey into space in 1961. Of course these achievements were overshadowed by NASA's landing of the first men on the Moon in 1969.

To facilitate future lunar landings, Russia is also preparing superheavy launchers. The Russian program will culminate sometime in the 2030s with crewed flights to the Moon. If the early results from the Indian Space Research Organization (ISRO) and from NASA on ice deposits are promising, this could be an initial step toward developing an ice-mining facility near the lunar south pole. Such an installation could provide a source of water and of hydrogen and oxygen with which to make rocket fuel.

It's not just the major space powers that are planning to robotically explore the Moon. Among the many robotic rovers planned for the early 2020s are missions from India and Japan. Even countries that have no launching capability of their own are involved in the lunar space race. On a smaller scale, Israel and the United Arab Emirates are joining in the challenge of the Moon as the emerging space agencies in those countries are intent on lofting spacecraft to the Moon via commercial providers. One example is the first privately funded lunar mission, the Israeli spacecraft Beresheet (or Genesis), which was launched on a SpaceX Falcon 9 rocket from Cape Canaveral. It crash-landed on the Moon in 2019. A successor mission planned for 2024 will include an orbiter and two lunar landers. One common goal of these missions is to measure the composition of soil samples from diverse regions.

The long-term aim is to develop bases and habitats on the Moon. Of course reaching this goal will take decades. If plans for an inhabited base are to succeed, choosing a location with abundant natural resources will be key. The preferred places currently envisaged for bases are near the lunar poles, where the climate is not extreme and water ice is most likely to be found in deep craters.

A new hydrogen economy beckons us to the Moon. The stakes are large, and the major rival space agencies are eager to establish priority. There are potentially enormous commercial benefits from lunar development. Tourist resorts would inevitably be developed in response to huge public interest. The developments under study range from tourist activities to mining and manufacturing under low gravity, and inevitably to rocket fuel depots and spaceports.

A combination of human and robotic assets will be needed to support science goals and should become integral to future lunar exploration. The will to pursue those goals is there, and the commercial aspects of lunar exploration are under intense study. I turn now to a discussion of the desirability of incorporating science into lunar exploration and its potential rewards for humanity.

Beyond the Moon

With the international space race underway, humans will certainly go to the Moon, whatever the cost. A lunar spaceport will serve as a gateway to the solar system for pursuing the long-term aim of exploring the solar system with crewed missions. Beyond the Moon, the primary target is Mars, which may be hiding clues to the origin of life, from a bygone era when its surface had flowing oceans. We need to dig deep on Mars to reveal clues to its past. Current robotic missions to Mars are planning to do precisely this, and the first major step is being taken by NASA's Perseverance rover. Perseverance landed in an ancient Martian river delta in February 2021 and will spend two years surveying the surface.

Of course the human aspiration to explore the solar system demands that we ultimately go beyond robotic missions. Mars is a challenging goal for crewed missions. Going there will inevitably be a risky endeavor. The Moon is only a three-day trip, and that path has already been navigated by the twelve Apollo astronauts who walked on the Moon, as well as by the six other astronauts who stayed in orbit on each landing. They too played indispensable roles in organizing the safe return to Earth.

Space planners are enthusiastic about the lunar sequel—a crewed trip to Mars. For the moment, a seven-month trip to

Mars, with a safe return, is currently beyond our capability for human space flight. Life support and indeed survival are among the unsolved problems. Exposure to galactic cosmic rays is one of the prime risks.[5]

Most cosmic rays are energetic protons capable of ionizing anything in their path. They are potential causes of gene mutations and cancers. To list just a few hazards of a trip to Mars, astronauts would be subject to vital organ deterioration, bone marrow damage, stem cell destruction, and tissue necrosis. Assisted only by current technology, astronauts would arrive on Mars saddled with muscular atrophy and riddled with cancers.

The dangers of radiation exposure depend on age and gender. The typical dosage is low on the Earth, where we are protected by the atmosphere. The average natural dosage per person corresponds to the biological effect of the deposit of some 30 ergs of X-rays in a kilogram of human tissue every year. One of the principal calibrations in risk estimation has been the survival rate of Japanese atom bomb survivors. The maximum limit by international standards is about 100 times the natural exposure limit. Adhering to this limit is required, for example, for individuals required to work with radioactive materials.

Our views have evolved on X-ray exposure. When I was a child, the best shoe stores in London routinely used X-ray machines for measuring shoe sizes. These delivered exposures of about one-tenth of the annual limit, typically in twenty seconds. And the wooden machines failed to protect the sales staff. Chest X-rays, by contrast, are well shielded and deliver typically about one-thirtieth of the natural limit.

The international regulatory limit is set by the estimate that cancer risk is doubled over a period of twenty years after exposure. Of course, such limits change with time and generally are lowered as more data become available. NASA's limit for low-Earth orbit trips to ISS is currently about five times higher than the current international limit on radioactive exposure, partly because astronauts are expected to take higher risks.

A recent report by the National Academies of Sciences, Engineering, and Medicine suggested that the lifetime maximum for a permissible dose should be reevaluated and recalculated to an even more conservative value for the highest lifetime risk to a healthy astronaut. The problem facing future space exploration is that a trip to Mars would greatly exceed this exposure. Meanwhile, traveling to the Moon should be safe.

Unshielded and prolonged space travel is dangerous, but most likely the risks will be manageable with enough spacecraft shielding and a correspondingly ultraheavy payload. The heavy-load spacecraft currently under development represent just the first steps toward developing the capacity that would be needed for a crewed Martian lander mission. When can we expect this to happen? NASA has announced plans to send a crewed mission to Mars by 2039. Such a mission would need to make use of lunar fuel resources. Launch of the necessary heavy payload would be achievable from a circumlunar space station.

Wealth in Asteroids

Meteorites and asteroids, which contain valuable natural resources that are rare or even nonexistent on Earth, have been bombarding the lunar surface over billions of years. The rain of cosmic debris was especially intense during the first few hundreds of millions of years after the Moon formed. The lunar regolith, at least 1 percent of which comes from ancient asteroids, is expected to be a rich source of rare elements. It promises to be a treasure trove for mining a century from now.

In the meantime, space agencies are exploring asteroids for extraction of samples to be returned to Earth. The study of asteroid surfaces provides a direct glimpse into the past, and asteroid mining is likely to be one of the long-term goals. Asteroids, which have no significant atmosphere, have much in common with the Moon. Asteroids and meteorites shaped the lunar surface over billions of years as meteoritic debris accumulated far more than on the surface of the Earth. In competition with ancient volcanic flows, asteroid debris covered the Moon. The buried layers lying deep under the lunar surface must contain huge amounts of asteroid debris.

Asteroid missions are expected to set the scene for a better understanding of the lunar composition. Heavy-load carriers are optimally launched from geostationary orbits rather than from the Earth. The low escape velocity to leave the Earth gives a huge fuel advantage. Launching facilities on such orbits are required to effectively mine asteroids and eventually tap the much larger resources of the Moon. It will be much easier to construct orbiting space stations using building materials brought in from low-gravity environments.

We know something about the composition of asteroid rocks because many meteorites are thought to be fragments of an asteroid parent body. The asteroid belt between Mars and Jupiter is a dangerous environment. Asteroids occasionally crash into each other and shatter, the debris dispersing throughout the solar system. Some debris reaches Earth, and the largest fragments survive the impact and are found as meteorites. But contamination by our atmosphere prevents us from having pristine samples of asteroid rock that survive impact. Going to an asteroid would allow the ultimate retrieval of rocks in space.[6]

Asteroid rock samples were first directly returned to Earth by the two Japanese Hayabusa missions. The minuscule samples collected during these missions were mere specks of rock, culminating with a 5-gram haul recovered when Hayabusa 2 landed in the Australian outback in late 2020. A more substantial sample was gathered by NASA's OSIRIS-REx spacecraft from an asteroid known as 101955 Bennu. A near-Earth asteroid discovered in 1999, Bennu is named after the ancient Egyptian mythological bird associated with creation, rebirth, and the Sun. After a seven-billion-kilometer round trip lasting seven years, the NASA spacecraft is expected to land in Utah in September 2023 with some 60 grams of asteroid rock samples.

Understanding the feasibility and profitability of asteroid mining is one of the ultimate goals of rock sample return missions. Also beckoning, however, are lunar resources, which, because of eventual in-situ manufacturing possibilities, could be far more cost-effective.

Lunar Mining

The international space agencies envisage many commercial activities on the Moon. China is developing plans to mine the Moon, and other countries will not lag far behind. The lunar surface is likely to be a unique site for mining rare elements,[7] including rare earth elements such as europium. The terrestrial supply of some of these key elements is likely to be exhausted over the next hundreds or thousands of years.

The applications of rare earth elements are legion. The many industrial applications include the manufacture of superconductors, smartphones, electronics, flash drives, light bulbs, camera lenses, computers, electric vehicles, catalysts, magnetic resonance imaging, and high-power magnets, as well as clean energy technologies, wind turbine dynamos, medicine,

X-ray tomography, and cancer treatments. Some rare earth elements are crucial to military applications, including laser weapons, radar, and sonar.

Rare earth elements are mined on the Earth, but only through environmentally polluting operations. Indeed, mining rare earths is such a toxic process that extraction is highly restricted. Economically viable deposits are limited to certain areas of the Earth. China is the site of the largest accumulation of rare earths.

Indeed, China dominates the limited world supplies of rare earth elements. Total world resources are 140 million tons, with more than one-third of these resources in China. The United States has 13 million tons of rare earths. Brazil and Vietnam combined have a reserve comparable to China's but currently are minor producers. At current extraction rates, some key rare earths are projected to be exhausted on the Earth in less than 1,000 years. The reserves of many rare elements face exhaustion within 10,000 years. These are short timescales for the terrestrial future. Of course, not all rare earth elements are rare—nearly all of them are more abundant than gold. But the extractable supply of rare earths is limited.

We will need a supply of the crucial rare earth elements for millions of years. Of course, further into the future, though it's impossible to reliably predict, humanity will surely have discovered new technologies that are less rare earth–dependent. Between then and now, lunar resources may bridge the gap. Based on analysis of the Apollo lunar samples, lunar reserves of rare earths approach a trillion tons—or 10,000 times more than the terrestrial reserves.

In light of how central rare earth elements are to present and future technologies, it will be hard for mining companies to resist the challenge of lunar extraction. The potential rewards are enormous, and the supply virtually inexhaustible. We will not run down the lunar reserves for a very long time.

Lunar extraction will be achieved robotically, and the environment will be closely controlled. Lunar habitats will be enclosed areas located near the polar regions or in giant lava tubes. There will be few local inhabitants to worry about. The toxic by-products could be shipped to the nearest and most efficient giant incinerator—the Sun.

Mining water will also be a major lunar activity. Liquid oxygen and hydrogen are derived from breaking down ice, and a by-product will provide reservoirs of rocket fuel for Earth transportation use and beyond. Lunar resources can serve us for millions of years. They represent the future for our planet.

A Fusion Future

One of the more intriguing mining resources will be the isotope of helium that has a mass of three atomic units.[8] The prevalent type of helium, Helium-4, is rare on the Earth. Helium-3 is thought to be primordial, its formation having preceded the solar system. Small amounts are found in the Earth's mantle, above the core and below the crust. Helium-3 is in high demand on the Earth for its cryogenic properties. It is the coldest refrigerant that exists. Because a helium-3 refrigerator cools down to 300 millidegrees Kelvin, it is extremely useful for industrial and scientific applications that require very low temperatures.

Helium-3 has been widely exploited in astronomy to build extremely sensitive cold detectors. These instruments search for tiny fluctuations in the cosmic microwave background, itself a radiation field that is at the equivalent of 3 degrees Kelvin.

Mapping these fluctuations with an ultracold detector has transformed the science of cosmology.

Thermonuclear fusion is another futuristic application of helium-3. Deuterium and tritium are currently the fuels of choice for ongoing experiments in energy generation by thermonuclear fusion. These elements are abundant on the Earth. However, their fusion produces neutrons, which interact with the containing device and create high levels of radioactive contamination. In contrast, helium-3 is a relatively clean fuel that undergoes thermonuclear fusion without generating neutrons. It would not produce dangerous radioactive waste products when burning in a thermonuclear fusion reactor.

Scientists believe that, in the long term, helium-3 could replace tritium and deuterium as a thermonuclear fuel source. There is a major problem, however, to be resolved. Fusion of helium-3 involves combining nuclei of relatively high atomic mass and nuclear charge, as compared to the usual deuterium-tritium mix in mainstream fusion technology. To overcome the charge barrier between helium-3 nuclei requires a much higher temperature than is needed for burning deuterium and tritium. This challenge presents a serious technological barrier.

The essence of fusion energy is extracting more energy from fusion than what is put in. Our current approach to controlled thermonuclear fusion has still not attained a sustainable supply of energy. It is difficult. The engineering requirements for successful nuclear fusion are considered extreme compared to our current generation of nuclear fission reactors. The timescale for achieving the first generation of controlled fusion reactors remains highly uncertain. The most optimistic estimates are for significant energy production by 2035. Others would argue for midcentury as more realistic.

Sustainable fusion, however, seems inevitable, and when we achieve it, our energy supplies will be revolutionized. But it's still relatively dirty energy, at least in the reactor vicinity. Once the national power grids are supplied with fusion energy, the race will commence to establish the next generation of cleaner fusion reactors.

While helium-3 may indeed be the cleanest possible fuel for unlocking controlled fusion, it is expensive. The terrestrial price exceeds tens of millions of dollars per kilogram. And a kilogram is about the total annual production on the Earth, where it is collected as a decay product of tritium. Terrestrial reserves amount to just a few tons.

The lunar regolith is estimated to contain millions of tons of helium-3. Most of it is the result of the helium-rich solar wind impacting the lunar surface over billions of years. The outermost layer of the Sun is a dilute hot gas or plasma of charged particles that is heated by coronal eruptions driven by magnetic storms. The particles are protons, electrons, and helium nuclei. Solar gravity does not contain the hot plasma, and a wind is produced that travels to the Earth and beyond. The solar wind is rich in helium, reflecting the composition of the Sun, and does not impact the Earth, where our atmosphere shields us from it. On the Moon, heating large quantities of lunar regolith will release vast amounts of solar-produced helium-3.

Helium-3 fusion is clean and green as an energy source. It is futuristic, but is under intensive study. The helium-3 recapture strategy is being led by the head of the Chinese Lunar Exploration Program, Professor Ouyang Ziyuan. Commercial investors are eagerly standing by to reap the rewards. Helium-3 is expected to be the element of choice for future thermonuclear fusion reactors, albeit not until a century from now. Lunar mining could help justify and even subsidize the development of lunar bases and lunar science.

A Moon Village

The European Space Agency has announced plans to build a lunar village for commercial activities, which are expected to include tourism as well as construction and mining.[9] Construction is especially feasible near the lunar south pole. Here both abundant ice on crater floors and continuous sunlight on crater rims are available. The temperature extremes are moderate. There is sufficient regolith and water to fabricate glassy bricks with which to construct dwellings. One current design study features a four-story building situated in a dark crater in permanent shadow.

Ambitious technologies are under development. State-of-the-art industrial complexes are capable of fabricating complex materials and machines out of the ambient lunar materials available—not just lunar regolith and water but also various other elements obtained from the lunar mining industry. The use of telerobotics will be central to extracting lunar resources and utilizing them.

A large construction industry will be developed, mostly run robotically with human supervision. Three-dimensional printing facilities will produce much of the material needed for local construction. The low-gravity environment will facilitate new manufacturing technologies that might be especially relevant for the biomedical and pharmaceutical industries. Local lunar industry is key, as transport from Earth is costly and limited to relatively small loads.

As the focus of international lunar activities, the lunar village is intended to serve a number of goals, including a sustainable human presence and activity on the lunar surface. Multiple users will carry out multiple activities. For so many reasons, the Moon represents a prime choice for pursuing political, programmatic, technical, scientific, operational, economical, and inspirational aims.

We can look forward to an era when human spaceflight has been developed for economic development and international cooperation. A key element will be innovation that will inspire and educate the workforce of the future. It is expected that the emerging lunar community will be the catalyst of new alliances between the public and private sectors.

As yet, little attention has been given to the unique advantages of a lunar platform for studying the Universe. Lunar telescopes should be a key component of a future Moon village. The science goals will enable advances in planetary science and our understanding of the origin of the Moon, and the astronomy goals will include the imaging of distant planets and the first stars. On this new frontier of exploration, we will probe the dark ages of the Universe, just as geologists study the origins of progressively older layers of rocks on Earth. We will do the same in space, where we will see that our "rocks" are the remote hydrogen clouds from which galaxies were assembled.

The lunar platform will facilitate novel applications of the life sciences so that we can better understand the importance of biological risk issues. If humans are to ever attain quasi-immortality, the lunar low-gravity environment will play a crucial role in developing the essential medical transplants. We will learn to replace all of our vital organs. As computers become ever more powerful, we may replace our brains as well—or at least upgrade their memory contents.

A new science vision needs to be implemented at this early stage of lunar planning. The possible discoveries from lunar exploration make an irrefutable case for science-driven projects in tandem with commercial activities. Although the timescales are decades or more, this should not hold us back from developing visions of the future.

Lunar Sites

Likely sites for the first sustainable human outposts include polar locations that provide moderate temperature conditions. There are many large craters at the poles, created by early impacts of asteroids. The advantage of the polar sites is that, as with the Earth's poles, the Sun is never very high in the sky there. The lunar poles are the ideal environment for avoiding extreme heat or cold. In the crater basins, the temperature is amazingly cold. NASA's Lunar Reconnaissance Orbiter (LRO) has produced infrared images that show temperatures as low as 30 degrees Kelvin. Volatile deposits form in the crater bowls, including ice.

The craters are typically several kilometers deep and up to hundreds of kilometers across. The rims are relics of ancient asteroid impacts that are high enough—up to four kilometers— to produce permanently shadowed areas. The crests of the crater rims are illuminated for up to 90 percent of the lunar year. Towers mounted on the crater rims could support solar panels to guarantee continuous solar power.

The dark polar craters, with a ready supply of solar power and water close by, are possible sites for developing lunar bases as well as for installing telescopes. The crater basins would be ideal locations for telescopes to perform high-resolution imaging throughout the lunar year. The low temperatures would facilitate construction of infrared telescopes that could probe the remotest regions of the Universe.

As discussed in more detail later, optical telescopes with very large apertures could be built in these dark polar craters, of sizes that are structurally inconceivable on the Earth. Imagine the ultimate hypertelescope with an aperture of kilometers—its resolving power would be superb. Construction of these telescopes would not be straightforward—the problem of ubiquitous and corrosive lunar dust particles would have to be resolved—but the payoff would be an ability to probe for signatures of biological activity in the atmospheres of many Earth-like planets around distant stars.

Giant Lava Tubes

There are hundreds of naturally empty lava tubes on the lunar surface, relics of a complex history of volcanic activity that covered much of the lunar surface in basaltic rock.[11] They have estimated widths of over a kilometer, lengths of up to several kilometers, and really high roofs. These tunnels were left over after episodes of volcanism melted the rock. The caves were drained by lava streams pouring under the cooling and hardening lunar crust. The basaltic lava cooled and flowed to lower levels. Giant empty caves or tubes remained.

The likelihood of there being giant lava caves in the lunar crust was discovered by orbiting probes of the lunar surface. In 2008, the Japanese Kaguya spacecraft discovered an opening to a giant lava tube in the Marius Hills region. NASA's Lunar Reconnaissance Orbiter followed this up in 2011 by imaging a 65-meter wide opening to a pit some 36 meters deep. The presence of lava tubes is inferred directly by such circular holes or skylights. The skylights are sections where the lava roof has collapsed. LRO imaged hundreds of similar skylights in subsurface caverns.

The NASA lunar science mission known as GRAIL (Gravity Recovery and Interior Laboratory) was launched in 2011 to do precision gravity mapping of the lunar surface. Two simultaneously deployed spacecraft used telemetry to measure their precise separations to within a micrometer. This enabled detailed mapping of the lunar gravitational field. The lunar crust was found to be far more porous than had previously been suspected. The presence of many lunar lava tubes up to a kilometer in width was inferred. If these tubes are several hundred meters deep, then the natural roofs of lava are expected to be at least a few meters thick. The lava ceilings provide a natural shield against deadly giant solar flares of energetic particles as well as against cosmic rays, small meteorites, and ejecta from asteroid impacts.

Shielded from potentially hazardous radiation, lava tubes are also protected against corrosive lunar dust, making them potential sites for human habitats. Some are large enough to provide space and shade for entire cities. The giant lava tubes are in diverse locations on the lunar surface but are typically found between lunar highlands and maria. These are strategic locations with easy access to landing sites and mineral resources.

By providing thermal insulation against the extreme daytime and nighttime temperatures, the deep caves are a place where the extreme temperatures of the fourteen terrestrial daylong lunar days and lunar nights—ranging from a freezing minus 130 degrees Centigrade to a boiling 120 degrees Centigrade—could be avoided. The ambient temperature below the basalt rock roofing—which may be up to 40 meters thick—would be stable throughout the year at about a cool minus 20 degrees Centigrade.

A biosphere will need to be developed for long-term habitation in the lava tubes, but there is already abundant water ice and a soil supply of oxygen. We can anticipate that the lunar construction industry will gear up to tackle the development of viable habitats as advanced robotic resources are implemented. This is a goal for the distant future.

Target: The Far Side of the Moon

The far side of the Moon provides a unique environment for low-frequency radio astronomy.[12] With no ionosphere and no terrestrial radio interference, it is the most radio-quiet environment in the inner solar system. Such an environment is needed to provide the exquisite sensitivity it will take to address the most challenging problem in cosmology—our origin. Only at very low radio frequencies can we hope to observe the cold gaseous building blocks of galaxies. Very early in our cosmic history, there was no light and there were no stars, but there were hydrogen clouds. Studying radio waves in the distant universe provides a unique way to probe the intervening dark ages.

Radio interference accompanies humans wherever they go, emanating from television transmission, cell phones, and marine radars and spilling over from many other terrestrial activities. Being shielded from the Earth, the far side of the Moon is a quiet radio environment, and the quietest part is in the middle of it. Terrestrial radio interference is reduced by a factor of 100 million compared to the near side of the Moon. This environment is the big quiet that will enable us to listen in to the Universe. Of course, it is uncomfortably hot and cold in such a location, but robotic deployment of a radio telescope should solve this problem.

A radio telescope whose resolution greatly exceeds that of any single dish could be built. One proposed design links an array of tens of thousands of radio antennae laid out over hundreds of kilometers. The deployment of an array of antennae could be optimized for sensitivity on the far side of the Moon. In this ultimate radio-quiet environment, where the Earth is never visible in the sky, all terrestrial sources of radio noise could be avoided.

We will still need to regulate radio interference from lunar activities, as we routinely do on the Earth by reserving certain frequency channels for radio astronomy. Low-frequency radio astronomy will open a pathway to cosmological investigation of the early universe.

Let's turn now to the basics. We have a great dream, but how do we implement it?

CHAPTER 2

Digging Deep on the Moon

Our enterprise, the exploration of nature's secrets, had no beginning and will have no end. Exploration is as natural an activity for human beings as conversation.

—FREEMAN DYSON

The Fossil Record

The surface of the Moon is a data recorder of the cosmic history of the solar system. Meteorites from throughout cosmic time have bombarded the lunar surface. Meteorites and asteroids proliferated early in the history of our solar system, like a gigantic storm that only settled down after millions of years, once the planets were formed. The result is a surface rich in meteoritic debris. Marked not only by splashes from billions of years of meteorite bombardments but also by extrusions from ancient tectonic activity and eruptive lava flows, the lunar surface carries a rich cosmic record, especially when we dig below the recent debris.

How did the Moon form? A leading theory invokes a giant impact with a Mars-size asteroid soon after the Earth formed. This long-vanished body has been named Theia. Its collision with the Earth raised immense amounts of rocky debris. The heavier debris settled in the backsplash of liquid magma and condensed in orbit to form the metallic core of the Moon. Denser minerals settled to form the Moon's rocky mantle, and lighter crystals floated to form the lunar crust. The young Moon orbited in close proximity to the parent Earth. The molten iron core of the Earth was largely intact. The lack of iron from the core of the Earth explains why the Moon is relatively iron-poor, unlike the core of the Earth.[1]

Not all is that simple, however. The chemical composition of the Moon is similar to the mantle of the Earth. Unfortunately, asteroids have differing compositions. So the impact of a giant asteroid cannot be the entire story. One suggestion is that Theia and the young Earth might have shared a common origin, and hence a similar composition.

Ever since its formation, the Moon has been drifting farther from the Earth. It is a remarkable coincidence that the Sun, far more distant and massive, exerts just about the same gravitational force on the Earth as the Moon does. Tidal forces exerted by the Moon act on the Earth in coordination with the tidal pull of the Sun, generating the twice-daily ocean tides. This same pull drags on the rotation of the Earth and causes the Moon to slowly drift away, but the total spin of the Earth and Moon system is unchanged. That's a basic law of physics.

The length of a terrestrial day used to be much shorter. The Moon formed when it was just a couple of Earth radii away, and a day on the Earth lasted only six hours. Now the Moon is 239,000 miles away.[2]

The outward drift provides support for the lunar origin in a massive collision of an asteroid with the Earth. The Moon formed from the splash-out debris of the Earth's mantle. The fact that its chemical composition is more similar to that of the Earth's mantle than to primitive meteorites is one step toward our theory of lunar formation—which, of course, is only a theory. A major goal of future lunar exploration will be to test this hypothesis with detailed measurements of the lunar composition in varying regions of the Moon and at different depths.

Cosmic Avalanches

A profusion of young stars, including our Sun, are demarcated by the spiral pattern of the Milky Way galaxy. The spiral arms that we see in many star-forming galaxies are like density compression waves. As the varying force of gravity, perhaps from an orbiting satellite galaxy, overwhelms the orbiting gas, the clouds are squeezed together. Cloud coalescences and compressions induce stars to form, and a spiral pattern of star formation is induced. Inner waves do not have as far to go and overtake the outer ones. Spiral arms are where glowing clusters of stars form contagiously like beads on a string. The spiral waves continuously renew themselves, thanks to the pull of gravity from a nearby neighbor. The Large Magellanic Cloud galaxy is our neighbor that drives our spiral arms.

Two hundred million years ago, young gas clouds completed their first orbits around the Milky Way. As they orbited they grew, sweeping up many smaller gas clouds along the path. Eventually the accumulation of mass was irresistible. Collapse was triggered as the cloud was compressed on its passage through a spiral wave. In the cosmic context, the giant snowball accumulated so much mass that it contracted under its own weight. Its own gravity prevailed. From gravitational collapse, stars were borne.

The precursor cloud to our solar system was very cold. About 1 percent of its mass was ice. That is our best estimate of the mass of the icy reservoir of future comets in the outermost solar system, light-years from the Sun. The nucleus of a comet is a block of ancient ice, perhaps up to 30 kilometers across. Comets swirled around our young Sun. Many comets that were on orbits that kept them far out were able to survive. Eventually, orbital perturbations, perhaps with Jupiter, kicked them inward to eventually become visible. Solar heating expelled steam, dust, and gaseous debris to form the beautiful tails of plasma and dust that trail behind a comet over millions of kilometers. Comets are the most primitive objects found in the solar system.

The Sun formed about 4.6 billion years ago in the center of a cloud. But it was not alone—a number of nearby stars formed at the same time. These are our neighbors in space, and many of them have their own systems of asteroids and planets. Proxima Centauri, the nearest star to us, is four light-years away.

Here is how our solar system formed. Thanks to the spin of the parent cloud, some of the gas, dust, and ice particles were retained in a disklike configuration around the forming star. Ices coated the dust particles. The particles coagulated into icy clumps with rocky cores. Many clumps in turn aggregated to form hordes of kilometer-scale icy rocks that we call planetesimals. These are the hypothetical building blocks of the planets. Planetesimals resemble giant dirty snowballs, with deadly cores of rock. Like snowballs, many of these ice-coated rocks agglomerate, like a giant block of ice rolling down a steep snowfield and building up in mass. An avalanche occurs as planets build up their mass from accumulating many planetesimals, along with small, icy dust particles and pebbles. And so the solar system was born. This is our best story, and it's one that accounts for the data.

There was a real storm of debris while the solar system formed. The massive icy planets (Jupiter, Saturn, Uranus, and Neptune) accumulated in the outer regions of the solar nebula, far from the emergent Sun. Rocky planets (Mercury, Earth, Venus, and Mars) formed closer in. Thanks to the local heating by the central star, the ices melted in the inner regions and the rocky objects predominated. Leftover debris formed icy moons around the giant planets. Uranus is an example of a collision with an Earth-mass protoplanet most likely tipping a giant planet on its side. The smaller gas giants Uranus and Neptune probably formed closer in to Jupiter's orbit, where there was a large disk of rocky and icy debris. While most of the debris was flung inward, they formed as they were flung out by repeated scatterings with smaller objects. Pluto was a remote icy relic, more of a moon than a planet. Farther out, the interplanetary survivors from the outer solar system are primitive meteorites and asteroids. Many of these were scattered along highly elongated orbits directed out of the disk. Their orbits enabled them to periodically plunge into the inner solar system.

Of course, we have no direct testimony to any of the events in this reconstruction of the past, but the solar system contains many clues that support its veracity. Reconstructing the birth of the solar system is rather like reconstructing the scene of a crime. One imagines that the planet formation sequence of mass accumulation culminated as huge dirty snowballs aggregated in a dusty disklike accretion belt around the young Sun. Some orbiting disks of rocky debris survive as failed moons in orbit around the gas giant planets—late-time relics of the past.

One example is Saturn's rings, frozen in time and orbitally directed by the shepherding of its many moons. They give us a glimpse of what the early solar system might have resembled in its early stages. Each of the giant gas-dominated planets, Jupiter, Saturn, Neptune, and Uranus, formed with a tiny, rocky, Earth-like core while retaining the light volatiles that formed methane-rich ices, as did their moons. Jupiter has a relic system of massive rings of rocks. Even Uranus and Neptune have systems of faint relic rings.

The First Clouds

Today star formation is very efficient as well as quite different from what must have occurred in the early Universe.[3] Gas cooling is so much easier now thanks to the omnipresent dust in the interstellar medium. Cooling in nearby interstellar gas clouds is aided and abetted by atoms heavier than hydrogen. In the coldest clouds, these atoms are essentially in the form of dirtlike specks of carbon and quartz. Clouds cool to low temperatures, tens of degrees Kelvin. Clouds fragment. Stars form prolifically. The typical mass of most stars that form today in a typically cold environment is close to the mass of the Sun. Very few stars are much more massive, but it is these that continue to die in explosions and contaminate the environment. Yet it was different in the early Universe.

The difference in cooling conditions distinguishes the primordial from the conventional star-forming environment. Carbon was not present in the first clouds because it had not yet formed. Carbon accumulates in the cores of stars by thermonuclear burning of hydrogen and helium over millions of years. It takes star deaths to make and liberate carbon. Long ago, star formation was a rare phenomenon, and star deaths were even rarer. The early Universe had no contamination from the heavier elements.

Clouds consisted predominantly of atomic hydrogen. There were a few leftover electrons from the early Universe. A hydrogen atom would catch a free electron to eventually form a molecule of hydrogen. None of the usual interstellar gas coolants, such as carbon, were available. But hydrogen atoms are not good coolants. They lack any low-lying energy levels to convert heat into radiation. Hydrogen molecules are a better bet, as they have lower energy levels than atoms, but even they are inefficient, cooling the first clouds down to only about 1,000 degrees Kelvin. That's so much more than the temperature of conventional star-forming clouds, just tens of degrees Kelvin.

Still, the key consequence was the loss of energy. The system cooled, the cloud contracted, and there was a tug of war between the gravity compressing the gas and the heat of the molecules providing gas pressure. Gravity won. As the cloud got denser and denser, it broke up into fragments. But fragmentation was not very efficient. Only a sprinkling of hydrogen molecules was around to radiate energy away, and more heat was retained, boosting the masses of the fragments. Massive stars formed but were destined to rapidly die in spectacular explosions.

The First Stars

Thanks to all this extra heat, the first stars were mostly hundreds of times more massive than today's typical stars.[4] Such massive stars are short-lived. They shine brightly and then die rapidly. Within a million years, their fuel supply is exhausted. Hydrogen first burns into helium, then helium into carbon, as the temperature rises in the core of the star. At the end of its nuclear-burning life, the star consists of the products of the nuclear reactions that powered its short existence—layers of helium, carbon, oxygen, and silicon. An outer coat of hydrogen surrounds the successive onionlike shells of enriched heavier elements. The end product of nuclear fusion burns the silicon and leaves behind a central core of iron, the most stable of the elements.

Stars implode when the central nuclear fuel reservoir has been depleted. That's when the iron core forms. The center of all stars that are more than a few times the mass of the Sun will implode, releasing huge amounts of energy that drives out the outer enriched layers and explosively ejecting highly polluted material. The explosion is seen as an incredibly luminous but transient star, a supernova. A hugely energetic object, a supernova is brighter than a million suns, but it is short-lived. A supernova fades within a year of its maximum light. Think of it as a participant, along with other dying stars, in a cosmic fireworks display.

The chemically enriched stellar debris rapidly expands. The ashes of the first stars contain large amounts of heavy elements, such as iron, neon, oxygen, silicon, and magnesium, that are spewed out into nearby space and end up in newly formed gas clouds. Here they are recycled into new generations of stars and eventually into planets. Future generations of stars will be loaded with the ashes of their progenitors. Supernova debris is the primary source of heavy elements in the Universe.

The next generation of stars to form includes stars of the mass of the Sun. With successive generations, the level of chemical enrichment builds up. Unlike their predecessors, most of the new stars are of lower mass and long-lived. The oldest generation of surviving stars is distinguishable from the many other stars in our vicinity by their extreme paucity of heavy elements. They are survivors bearing testimony to their long-dead parents. They are living witnesses of the beginning.

With a new generation of huge lunar telescopes, we can hope to directly image the first stars—the luminous and short-lived stars that eventually explode. We will be looking back in time, digging deep into the Universe, to catch the faintest glimmers of star formation and death at the end of the dark ages.

The History of Hydrogen

The hydrogen atoms began to form 380,000 years after the beginning of the Universe, when the electrons and protons combined. Before this, the Universe was hotter, hydrogen was ionized, and matter was dense enough for the free electrons to cause many scatterings of the background radiation photons. That's how far back we can see with the cosmic microwave background radiation. It's only after a million years have elapsed that the Universe is purely atomic. Later, we enter the dark ages. The Universe stays dark until 100 million years elapse, when the dark ages end with the formation of the first stars.

We can explore the dark ages, going back to the time before there are any stars. With no stars, there is no light. But there are hydrogen atoms. The epoch of atomic hydrogen is a window in time between one million and 100 million years after the Big Bang. This provides our best opportunity to directly study the beginning. The imprint of gravity on these hydrogen clouds is unimpeded by the complications of stars. The clouds offer a clean view of the past. By studying the many shadows they cast in absorption against the background radiation, we can use the power of radio astronomy to map these shadows. We can unravel the details of the earliest imprints that seeded the growth of structure. The dark ages highlight the appearance of the first clouds and set the scene for the formation of the first stars.

We have a cosmic screen that catches fossil signatures from the beginning with unprecedented accuracy, capturing brute force with extreme precision. The resolution is much finer than can be achieved by surveying billions of galaxies to trace out large-scale structure. It's like going from using a hammer to dissect a fossil to using a molecular forceps. For the clouds, the increase in resolution is a millionfold. And we will achieve these images best from the Moon.

A Bottom-Up Process

The first small galaxies grew thanks to the pull of gravity in a process that we describe as gravitational instability.[5] Gravity is the most powerful force in the Universe. There is no resisting gravity once it dominates. The first clouds were drawn together by their own mutual gravitational attraction as the Universe expanded. Gravity prevails locally against the expansion of the Universe. The clouds fell together into larger and larger aggregates. The clouds were naturally cold, and many clouds merged together.

Matter continues to pile up. Normally, it is gas pressure that limits cloud growth. In nearby interstellar space, this is not a problem. The heat is eventually radiated away, and pressure is reduced. Clouds are not limited by acquiring too much heat and associated thermal pressure. Gravity prevails. Clouds merge together and grow in mass as more and more gas accumulates. Structure forms from the bottom up.

Larger and larger masses condense out of the expanding universe. Massive halos such as that of our Milky Way galaxy assemble from a hierarchy of many smaller merging clouds. Some memory of subhalos remains as the massive halos are forged by gravity. Computer simulations have confirmed this picture of bottom-up formation of structure. Some of the original clouds indeed do not merge but eventually form an assembly of leftover small galaxies, the dwarf galaxies. Each weighs in at millions of solar masses or more. We predict that there should be large numbers of dwarf galaxies for every galaxy like the Milky Way galaxy. According to the simulations, there were

once millions. Now, however, only thousands remain, according to the observations. We need to understand this shortfall.

The masses of stellar survivors are set by the requirement
that gas clouds have enough gravity to overcome the gas pressure. The gas cools and contracts to form what eventually turns
into a star-forming cloud. This is a dwarf galaxy. Smaller clouds
don't survive to the end because their gravity is too weak. They
are failed dwarfs. This accounts for some reduction in numbers.
Others are disrupted by encounters with the denser parts of
galaxies.

There are relic traces of the long-vanished population of
small galaxies. As the dwarfs dissolved to form our galaxy, they
left behind fossil signatures of stellar debris. These are seen as
long stellar trails from long-disrupted small galaxies. The trails
consist of ejected stars that maintain spatial patterns over several orbits. Our surrounding halo of stars has persistent long-
lived dynamical relics from the formation of the Milky Way
galaxy. These are frozen-in fossils from the past.[6]

As we peer deeper into space, we see galaxies evolving. Once
upon a time, a few billion years ago, they were mostly young. In
the past, young galaxies had a much higher gas content and
formed stars prolifically. We see vigorous episodes of star formation long ago. Distant galaxies were often much brighter
than they are today.

The millions of galaxies throughout the Universe that we can
survey slowly exhaust their gas supply, the raw material needed
for star formation, as the galaxy gas content is gradually depleted. The history of galactic star formation reveals a peak at
about four billion years after the Big Bang. That is also when the
Milky Way galaxy formed, about 10 billion years ago. We are
part of the crowd, just a typical galaxy, with stellar memories of
our past glory.

The Dark Side

About 85 percent of the matter is not atomic hydrogen and helium, the stuff of which the stars are made, but rather weakly interacting elementary particles, the so-called dark matter. The ordinary matter scatters with the radiation. It warms up and attains the same temperature. The dark matter does not scatter with radiation. It stays cold.

Dark matter is indeed cold. If matter is cold and its gravity is important, then it is intrinsically unstable. Cold matter fragments. Dark substructures are produced on all scales. Only a small fraction of these structures contain enough ordinary matter to form stars, which are visible as small galaxies. For this to occur, loss of energy by cooling is required and high gas densities must be achieved. Dark matter cannot cool. It is left behind and does not attain high densities. Dark matter does not form stars. The ordinary matter merges seamlessly together as it forms stars. A dense stellar system is formed that we recognize as a luminous galaxy.

We expect that the Milky Way is itself surrounded by an immense halo of dark matter. Even though it emits no light, the halo can actually be traced out by astronomers, who measure the rotation of the galaxy. Suppose all of the galactic mass were concentrated where we see the visible light. Then the orbital velocities of stars and gas clouds should decrease with increasing distance from the center. A similar effect, the decrease of gravitational force with distance, is predicted by Newton's theory of gravity. Similar reasoning long ago led Johannes Kepler to demonstrate that gravity could account for the motions of the planets around the Sun.

A different story emerges for the motions of stars and clouds in our galaxy and in many other galaxies. Modern observations that measure the velocities of luminous stars in the outskirts of the Milky Way have detected no decrease. There seems to be missing mass, a shortfall that would require a new theory of gravity. Or perhaps there is unseen dark mass. We'll see that there is increasing evidence for dark matter. This interpretation fits best into our understanding of gravity.

A pioneer of dark matter was Vera Rubin. One of the first modern female astronomers, she worked in the 1960s at the Mount Wilson Observatory in Pasadena. This was at a time when Mount Wilson was very much a male-dominated facility. There were few female astronomers, and they were generally dissuaded from observing with the great California telescopes. Rubin persevered. She recounted that one method of discouraging female astronomers was to provide no facilities for females on the mountaintop. Indeed, this limitation was stated clearly on the proposal for telescope observing time.

Rubin targeted emission from optical nebulae. The atoms in these gas clouds were excited by massive stars around the Andromeda galaxy. Andromeda is our nearest galactic neighbor that is close to being a Milky Way counterpart. The emission included peaks at discrete frequencies or spectral lines that corresponded to the excited energy levels of the gas atoms. Rubin was able to measure the Doppler shifts due to the rotation of the gas nebulae around the galaxy. Her work clearly demonstrated that a non-luminous component of matter was responsible for "flattening" the measured rotation curve. The star velocities did not decrease with increasing distance from the galaxy, as expected if gravity traced the distribution of starlight. But there remained some ambiguity. Perhaps the outer stars were intrinsically dimmer and being missed. One had to go farther out, to the outermost regions of the Andromeda galaxy, to be sure.

Using a new technique, astronomers soon peered further into the dimmest regions of Andromeda where there are few, if any, visible stars. But stars were not needed. Targeting clouds of atomic gas, radio astronomer Morton Roberts measured the hydrogen clouds that extended out beyond the stars to much greater distances. The emission of atomic hydrogen was used to probe the orbital velocities of the clouds in the outskirts of the Andromeda galaxy.

The astronomers discovered that the clouds were orbiting Andromeda in nearly circular orbits. As they looked farther and farther away from the bright galaxy, however, the orbital speeds did not decrease. They stayed constant where there were essentially no stars. The extra push felt by the clouds could not be due to the cumulative effects of the gravity of the known stars. Something else was at work that added to gravity.

This could only mean that invisible or dark matter was present. Dark matter is needed to provide the extra gravitational force that explained their velocities. The dark matter required amounted to ten times the mass in luminous matter. The dark stuff extended to ten times the radius of the luminous matter from the center of the galaxy. Of course, it might be that Einstein's gravity theory breaks down in the outermost regions of galaxies. But this more radical possibility has received scant support.

Following the pioneering maps of the Andromeda galaxy, it was not long before dark matter was found in many galaxies, and in particular close to home, throughout our Milky Way galaxy.[7] So it's not the gravity of the stars that helps to maintain their orbital velocities. This gravity is too weak. It's the dark matter.

The Universe in a Computer

Cosmologists like to put galaxy formation into a computing program that evolves the Universe forward in time. We can make a simulation of the formation of structure, and we can do this so well that a simulated galaxy looks like the real thing. This similarity does not guarantee the accuracy of the simulation, but the simulation is a useful guide nonetheless. It lacks key details because we cannot match the dynamical range that we observe from stars to galaxies. Numerically obtaining the required resolution is difficult. But these simulations are part of a rapidly developing field as computers become ever more powerful.

Indeed, a big advance in our understanding of the distribution of dark matter came from computer simulations, which enabled astronomers to confirm that dark matter is the dominant constituent of matter. The gravity of dark matter was imprinted on the large-scale distribution of galaxies. Without dark matter, the simulations could not fit the data.

Dark matter's gravity supported the giant clouds that preceded the galaxies. Within the clouds, ordinary matter fragments and condenses into stars. The stuff that forms the stars is mostly hydrogen and helium, what we collectively refer to as ordinary, baryonic matter. The galaxy is rotating. As matter aggregates, it conserves its angular momentum. A rotating disk of baryonic matter forms within a surrounding dark halo. A central spheroidal bulge of stars also forms. We have constructed the Milky Way galaxy in a computer.

Surprisingly, the dark halo turns out not to be a gigantic amorphous blob. Rather, it is teeming with dark matter substructures. The dark matter components—the dark halos around galaxies—retain signs of their primordial substructure. It is as though there is a memory of the assembly of the halo, which formed through the merging of many smaller dark clumps. The computer models reveal many thousands of smaller dwarf halos orbiting within the parent halo.

The Milky Way in fact has a stellar disk and a bulge of stars. The disk is fairly thin, and the bulge is fat and slightly prolate. The simulations recover these aspects, but there is a problem in the outermost parts. The surrounding halo shows something unexpected. We observe hundreds of dwarf galaxies, not the expected tens of thousands. Are our simulations realistic?

The numerical simulations are only as good as the physics that we use in setting up a simulation. Getting the physics right turns out to be remarkably difficult. Much depends on our detailed understanding of how stars formed, but we struggle with this in the Milky Way galaxy. Star-forming sites are deeply embedded in dust, stirred up by the turbulence generated by young stars, and threaded by magnetic fields. It's a heady mixture that astronomers are still striving to understand.

There is a further complication. Our observations of star formation are fairly local. Different physical conditions are likely to have prevailed in the early Universe, but it's far away. Here our observations are at much lower resolution and inevitably less detailed. Throw in our current inability to simulate the scales that span the wide range from planets and stars to galaxies and the outcome is a great uncertainty in our modeling of star formation. And it gets more complicated.

What happened to the predicted satellites? The difference is likely due to the fact that dwarf galaxies are extraordinarily inefficient at forming stars. Many are too starless and too dark to even show up in the surveys. It takes only the explosion of an early supernova in one of these tiny forming dwarf galaxies to blast out most of the residual interstellar gas. Not much is left—just a few stars that formed earlier.[8]

And as if that were not enough, there are further disasters awaiting any survivors. Dwarf galaxies have low escape velocities, that is, it takes only a small kick for stars to be ejected. Passage through the galaxy disk would provide such impulses by gravitational interactions. Stars will receive kicks as the dwarfs pass by stellar associations. So we infer that many dwarfs were tidally disrupted.

We saw earlier that patterns and tidal tails in halo stars trace long-disrupted dwarfs. Some of these early dwarfs might have barely survived, retaining a few stars from their initial formation phase. They would be exceedingly faint but in principle are detectable. Our models predict that there are likely to be hundreds of really dim dwarfs in the outskirts of the Milky Way. These are the last survivors.

Sophisticated telescopes have been developed that specialize in searching for very faint nearby galaxies. Some are being detected as ultra-diffuse dwarf galaxies, barely visible against the night sky brightness. The harder we look, the more we find. So the pieces of the cosmic growth puzzle are slowly coming together.

Pollution

Cosmology has taken another fruitful direction.[9] The surviving relics, dwarfs and very old stars, satisfy the expected property of being chemically primitive. They formed long ago, before successive generations of stars polluted and enriched the interstellar medium. Old stars have very low amounts of heavier elements. Comparison of the spectra of the oldest stars in the dwarfs to those of stars in the disk reveals their age. The oldest disk stars formed billions of years after their dwarf counterparts. They are relative latecomers. Many dwarfs are really old.

Heavy elements are synthesized by successive generations of stars. After massive stars explode, their debris is ejected into the surroundings of the dwarf galaxies. As time goes on, larger galaxies form from the cumulative merging and debris of the smaller dwarfs. A massive galaxy retains the stellar pollutants in its interstellar medium. The gas is recycled into new stars. Systematically more enriched stars are formed. The Milky Way is an amalgam of many smaller mergers.

We can unravel the chemical history of our galaxy by studying the distribution of the various chemical elements abundantly found in different regions. We do this for stars of different ages. Stars are like chronometers. Their chemistry is a clue to their age. We study chemical evolution by looking at the temporal and spatial variations in the relative amounts of different elements.

Iron is the simplest element to study. It is the most direct monitor of age and is synthesized in supernovae that exploded long ago. The iron abundance builds up with time as more and more massive stars explode. Hydrogen is the major constituent of stars. The less iron one finds relative to the hydrogen, the older the stars are inferred to be. We are looking back in time over several billion years. The first Sun-like stars in the galaxy are only one-millionth as iron-rich as the Sun. These metal-poor stars formed five billion years before the Sun. They are old.

Their planets are equally old, raising a fascinating question. Has life had enough time to evolve on old Earth-like planets? How much enrichment by heavier elements is needed to favor biological evolution? Billions of years are available. We return to this question later when we establish goals for lunar telescopes.

Grown Up at Last

The first phase of galaxy formation happened just 100 million years after the beginning.

There is eventually an end to galaxy growth. As the mass of the parent cloud destined to form a galaxy increases, the pull of gravity becomes more intense. The galaxy-forming cloud heats up.

FIGURE 2. A global view of Earth at night. NASA researchers have used these images of nighttime lights, compiled from over 400 satellite images, to study weather around urban areas. Galaxies too are sparse; the great agglomerations dominate.

Image credit: NASA Earth Observatory/NOAA NGDC, 2012, https://www.nasa .gov/sites/default/files/images/712130main_8246931247_e60f3c09fb_o.jpg.

The forming galaxy risks becoming so hot that continuing radiation of energy is inefficient. This occurs at a mass of about ten times that of the Milky Way. Such galaxies would be the most massive galaxies. No more massive galaxies can form.

The big galaxies are often surrounded by lesser galaxies. The entire ensembles are groupings or clusters of galaxies. The spaces between the galaxies contain intergalactic gas, much of which has been blown out of galaxies. Most of the gas is left over from when the galaxies formed. Owing to the enhanced gravity available in a cluster of many hundreds of galaxies, the intergalactic gas heats up and glows in X-rays. With X-ray telescopes, we routinely observe thousands of giant clusters of galaxies in distant regions of the Universe.

The galaxies orbit a massive galaxy cluster at high speeds, often moving at thousands of kilometers per second. They punch into hot intergalactic gas that acts like a battering ram on the cold interstellar clouds in the galaxy. The galaxies are stripped, especially those in the inner parts of the cluster. This is the origin of most of the intergalactic gas. The gas contains clues to the chemical history of the galaxies. Intergalactic gas is enriched with heavy elements. Some heavy elements ended up in stars, and some did not. The leftover gas in between the galaxies is a huge chemical reservoir that provides evidence about the history of the first stars that synthesized much of the iron in the Universe.

Thousands of galaxies aggregate into the greatest clusters. Imagine a crowded football stadium. There are only a few stadiums in a big city, but they are packed. The density of fans greatly exceeds the typical density of people in the streets. Proximity is the key.

Clusters of galaxies are relatively sparse. The nearest great cluster is tens of millions of light-years away from us. Yet the proximity of galaxies in a cluster is hundreds of times higher than for the typical galaxy in a low-density environment. Clusters are a crowded environment. Indeed, galaxy collisions are common and leave telltale distortions in galaxy morphologies. Gravity is the glue that attracts and retains galaxies in clusters. Telescopes in lunar observatories will help us unravel the hidden secrets of the formation of the first stars and the assembling of the galaxies as the dark ages drew to a close.

CHAPTER 3

Robots and Humans

We can follow our dreams to distant stars, living and working in space for peaceful, economic, and scientific gain.

—RONALD REAGAN

Robots would do a better job and be much cheaper because you don't have to bring them back.

—STEPHEN W. HAWKING

Space Hazards

Space is a dangerous environment, and the Moon is especially vulnerable. There is no atmosphere to protect the surface against extreme solar flares and coronal mass ejections. Solar energetic charged particles are deflected by the Earth's magnetic field. There is also bombardment by micrometeoroids that burn up in the Earth's atmosphere. The Moon lacks this safety net.

There are potential dangers awaiting astronauts on the lunar surface. The exposure to energetic particles capable of causing cell damage is at least a hundred times higher than on the Earth. And rare energetic events can greatly augment this risk. Occasional flares from the solar corona can boost the cosmic ray flux to extreme levels for periods of up to a few days. Even more dangerous giant solar flares often occur near the peak of the eleven-year solar activity cycle. The high-energy particles from solar flares are highly penetrating.

Our galaxy produces high-energy cosmic rays. They are everywhere. On the Earth our atmosphere buffers the surface from high-energy particles, providing a comfort zone that mitigates exposure to the associated risks. This does not happen in space. In contrast to the risks to humans of the long voyage to Mars, astronauts arrive safely after the short trip to the Moon. But once they have arrived, their exposure to high-energy cosmic rays must be limited.

Much rarer and more powerful events occur over centuries from giant explosions in the solar corona that are potentially even more destructive, although the particles are less penetrating than galactic cosmic rays. Advance warning is the key for astronaut survival, since the propagation time from the Sun is about fifteen hours for the fastest-moving solar eruptions. Enhanced protection for astronauts on the lunar surface will be achievable with cosmic ray storm shelters, using plastics or water tanks. The best solution for protection against high-energy cosmic rays may involve habitations within the giant lava caves.

The surface of the Earth is a relatively safe environment. Occasionally we find meteorites in the desert or on land surfaces where they may be more easily be detected. The location of the impact is often key to finding a meteorite. A hiker is very unlikely to be subject to a cosmic impact. The larger meteorites usually leave fragments on impact. Some meteorites survive until immediately prior to potential impact but explode above the surface, sometimes leaving no sizable relic or impact crater.

The most powerful meteorite impact in the modern era was just such an explosive event. Presumably the meteorite was mostly ice and lacked a rocky core. The Tunguska event occurred in eastern Siberia in the morning of June 30, 1908, and is thought to have arisen from the impact of a meteorite about 100 meters in size. This would be the equivalent of a small asteroid. The impact had the energy of a thermonuclear bomb, or some 10 megatons. The explosion about 10 kilometers above the Earth's surface flattened some 80 million trees over 2,000 square kilometers. Events such as the Tunguska airburst are estimated to occur every 1,000 years. Smaller air explosion events with atomic bomb energy (kilotons) are thought to occur yearly. The number of surface impacts is expected to be comparable.

A more recent massive airburst occurred in 2013 over Chelyabinsk near the Ural Mountains in Russia. It was caused by the largest asteroid known to have entered the Earth's atmosphere since the Tunguska explosion. The asteroid was some 20 meters across and exploded approximately 30 kilometers above the ground. Unlike Tunguska, many small meteoritic fragments were found. The energy involved was about thirty times that of the Hiroshima atomic bomb, and the light flare was more luminous than the Sun. There were no direct injuries to bystanders but the damage to buildings from the associated shock wave was considerable.

The Moon, where there is no protective atmosphere, presents quite a different story. The smallest meteorites survive impact. The risk for astronauts is significant during any extended activity on the lunar surface. Catastrophic impact events pose a real problem in light of the vast numbers of potential killer asteroids in the outer solar system. Asteroids are occasionally deflected toward the Moon as they pass near giant planets such as Jupiter. An early warning system will provide the required security by allowing time for evacuation to a safe environment. Anyone on the lunar surface might need to evacuate to underground caves or lava tubes for protection during extreme events.

From Habitats to Rocket Fuel

Protection against radiation risks needs to be incorporated into lunar habitat construction, which can be done using the raw materials available. A shelter with a thick roof of aluminum will protect against all solar cycle particle events. Lunar living quarters can be constructed as domelike structures, making optimal use of local materials. A very fine dust of lunar regolith can be combined with water to make cementlike bricks, providing the basis for developing a local brick industry. The bricks used in lunar structures will be designed to withstand damage from extreme temperatures and micrometeorite impacts.

But these measures will not suffice for withstanding meteorite impacts. Natural lava roofs will provide greater protection in giant lava tubes, which may be among the best sites for major lunar habitations. The best protection against meteorites is a thick protective roof. The dome roof will have to be at least a couple of meters thick just to insulate against the most penetrating energetic particles. Lava domes should provide a natural solution. They will protect against meteorites, although a rare very large meteorite impact could be destructive. Most meteorites have rocky cores and would shatter on impact, but even the debris would be dangerous for unprotected astronauts.

Much future lunar exploration will be performed robotically. The terrain and the environment are too hazardous for long exposures on the surface. But humans will need to direct this activity *in situ*. The human-robot interface presents interesting challenges in the execution of highly complex, long-time-line, and large-scale infrastructure tasks. Robots will be utilized to undertake lunar exploration, survey the lunar surface for potential landing and mining sites, take soil samples on the surface and subsurface, and analyze them in real time.[1]

Once bases are installed on the lunar surface, a key element of lunar development will be exploiting the water supply, both to make rocket fuel and for construction uses. Fortunately, the Moon has an abundance of water in its soil. Water ice deposits have been found in permanently dark polar craters. Water can also be extracted from lunar regolith, where, as mentioned earlier, the presence of water molecules has been discovered. Lunar resources are expected to provide an adequate supply of water. With water reservoirs available, the Moon can become a reservoir of natural propellants for interplanetary travel.

Rocket engines are powered by the chemical combustion of hydrogen and oxygen. A ton of liquid oxygen propellant can be generated from five tons of processed lunar water. The idea is to electrically dissociate the water into hydrogen and oxygen. The hydrogen and oxygen would be liquefied from the ice in the cold lunar craters, and refrigerants would be developed with energy from solar power. Both liquid hydrogen and liquid oxygen would be stored in gigantic lunar fuel stations.

Lunar mining will propel the new hydrogen economy. The Moon's low gravity will make it possible for the potential fuel supply to be developed for relatively low-cost spaceflight. There will be a virtually unlimited energy source for lunar-initiated space travel. Eventually, lunar spaceports will service travel throughout the solar system.

Robotic Deployment

The functionalities and capabilities of robots will be developed to allow control that is directly coupled with human-machine interaction capabilities.[2] Machine learning techniques and advanced artificial intelligence will enable robots to undertake complex autonomous functionalities. Humans will still need to be within reach.

Imagine robotic rovers scurrying around in the lunar desert. Direct human-telerobotic interfaces will benefit from advanced technologies to implement measures such as hazard avoidance and autonomy. Information will be generated for human guidance via augmented reality and visualization. Mobile manipulation robots will be equipped with complex robotic task planners to control navigation, object detection, and sophisticated communications with semantic interpretation and voice control. Human-supervised robotic operations need to be time-efficient in order to promptly cover interactions, manipulation, and handling capabilities. The end point is to allow deployment, operation, and maintenance by multiple robots supervised by a minimal number of humans.

Deployment control and robot management can be best achieved remotely from an orbital lunar space station. Mining sites will be located and operated far from residential sites. Human supervision will play a critical role. The communication time delay to Earth of about one and a quarter seconds makes local control essential. Autonomous robots will still need some local supervision for decision-making on-site. The construction of lunar bases will be a key phase in setting up the infrastructure to build even more grandiose projects in the future.

In exploiting the Moon's vast mineral resources, we should bear in mind that extraction can be a polluting and environmentally disfiguring process. Deposits of certain lunar minerals may be highly concentrated in some areas. For instance, deposits of thorium, a weakly radioactive metallic element that emits gamma rays, have been mapped by NASA's Lunar Prospector spacecraft near the lunar surface. These are likely sites of radioactive uranium and rare earth elements. When the Moon formed from a major impact long ago, the resulting melting produced an ocean of liquid magma. Lighter rocks, including those with enhanced thorium and rare earth elements, floated to the top and are found between the lunar mantle and the crust. Here they can be exposed by a major asteroid impact, which occurred in the South Pole–Aitken Basin, one of the largest far-side craters at 2,500 kilometers across. Previous thorium mapping had revealed enhanced deposits in the largest of the lunar maria, Oceanus Procellarum, which is some four million square kilometers in area. The lunar maria are dark, basalt-rich plains formed by ancient volcanic eruptions. Another thorium-rich site separated from Oceanus Procellarum by the Carpathian Mountains is another large lunar crater, the Imbrium Basin.

These are potentially prime mining sites. As mentioned earlier, the Chinese probe Change's collected the first rock samples from Oceanus Procellarum at the end of 2020. Future plans envisage mining more than metals. Lunar soil is easily excavated and rich in oxygen. As we have seen, water ice is abundant in permanently shadowed craters, and solar power is abundant.

It will be necessary to develop nonpolluting technology in order to avoid adverse impacts on scientific exploration and on human habitats. Autonomous robots will support much of the activity. One inevitable consequence of lunar exploitation will be the essential need to develop reconciliation procedures to help negotiate between competing or contiguous mineral claims. The legal implications of such claims are discussed in a later chapter.

The scientists should have their day. Robotic technology will be applied to site searches for radio and infrared telescopes. The scientific instruments will be maintained by robots, which will be integrated into modular transporters that enable observation and monitoring of the telemetric data. Data from astronomical measurements and geological surveys will interface with the mechanical, communication, and electrical facilities on the lunar surface. Progress should be rapid. The communication infrastructure for deployment and operations, and the interface with human supervisors, will be developed well beyond anything that we can currently envisage.

Lunar Dust

The lunar regolith is powdery gray dust that covers a thin layer of the lunar surface that is only a few meters thick. It is the result of billions of years of bombardment of the lunar surface by asteroids. As the surface particles are electrically charged by sunlight, they lose electrons and acquire a weak positive charge. The resulting electrostatic forces cause these dust particles to levitate as they become slightly repelled by the surface. But they are thought to settle down at the onset of lunar night.

Lunar dust consists mostly of silicon dioxide, glasslike particles with the consistency of flour. It adheres easily when disturbed, is highly abrasive, and poses dangers for both humans and machines. Lunar dust is not good for telescope operations. In the unfriendly environment of the lunar surface, protection will be needed from dust as well as from ionizing solar radiation and the solar energetic particle events. Telescopes will need careful maintenance, and the telescope mirrors installed in the dark lunar craters will need to be permanently protected.

The origin of the dust lies in a complex history of meteorite bombardment and ancient volcanic activity. Let's use the Earth as a guide. The terrestrial surface was shaped by highly energetic events, and we can study those of the past million years. Extreme weathering and tectonic activity hide much of the earlier record. The lunar surface is an easier book to read. Thanks to the lack of an atmosphere and flowing water, lunar soil has a much longer memory than terrestrial soil.

Lunar dust is everywhere. Its depth ranges from a few meters to several meters or more in the oldest areas of the lunar highlands. We have seen that lunar dust—which is similar to the volcanic ash in ancient terrestrial lava flows—can be combined with water to provide bricks as construction material. Plans include building roads and launching sites and constructing buildings. Another proposal is to manufacture polished glassy telescope mirrors out of lunar dust and water, bound together by epoxy.

Dust is rich in oxides, and oxygen could be extracted from it. There are many potential applications of this oxygen, such as producing fuel or even sourcing a biosphere in controlled environments such as lava tubes.

Tourism

There is a pent-up demand for tourist travel into space, and some of the lunar construction done by robots will focus on addressing that demand. In the twenty-second century, the breathtaking experience of watching Earth rise will become the apex of tourists' dreams.

Today's Grand Tour visits the top wonders of the world, including the Taj Mahal, the Great Wall of China, the Colosseum of Rome, the temples of Angkor Wat, the desert city of Petra in Jordan, Chichen Itza in the Yucatan, Machu Picchu in the Peruvian Andes, the Christ the Redeemer statue in Rio de Janeiro, the Eiffel Tower in Paris, and the great Pyramid of Cheops in Cairo. Humanity has marveled over these sights for many decades. But seeing Earth rise on the Moon will surely top them all.

The large hotel chains are salivating at the prospect of developing extensive resort complexes on the Moon. Leisure facilities could feature unique activities ranging from lunar golfing to quad rover rides through the regolith. Such activities will require more than just accommodations. There will be extensive planning for permanent installations. As in our greatest cities, culture and science will be incorporated into tourist activities. Peering into the crystal ball, there could be cultural centers, museums, scientific infrastructures, hospitals, sports complexes, transportation hubs, universities, and, inevitably, commercial establishments, from shopping centers to lunar factories and data centers.

This reads like science fiction, but I believe that it is where we will be before the century is out. Today's overwhelming demand for lunar tourism will be addressed initially with orbital trips around the Moon. Tickets are already being sold by the likes of SpaceX for launches planned within five years. Tourist lunar landings now seem pure fantasy, but the will is there. Such a hugely commercially driven activity, with vast potential returns, is attracting some of the wealthiest entrepreneurs on our planet, including Virgin Galactic's Richard Branson, Amazon/ Blue Origin's Jeff Bezos, and Tesla/SpaceX's Elon Musk. It seems inevitable that tourism will develop into one of the major lunar activities. Lunar rides, lunar walks, even lunar golf could be featured, echoing the activities of the early astronauts.

On a longer timescale, giant resort complexes could be developed with luxurious facilities for the super-rich that would never be achievable on Earth. Sports would include golf courses, a potentially exciting activity in a gravity field that is only 15 percent that of the Earth. A golf ball could be driven for up to two miles on a lunar golf course, the ball remaining aloft for nearly a minute. And space excursions would also be facilitated by the low lunar gravity. Imagine visiting an asteroid. Or even a comet.

No doubt the commercial sponsors would go first for luxury tourism. Here the profit margins are attractive, given the limited travel capacity expected in the early years of lunar travel. But imagine the development of large people carrier spacecraft. Eventually, mass-market tourism to the Moon could be developed. Perhaps lunar vacation packages would be pioneered by China, host to one of the world's most rapidly expanding tourism markets. There is much to be said for sharing lunar resources, and China already plans to be a leading lunar base developer. Mining might be China's main goal, but it seems inevitable that tourism would follow.

A Base for Science

Scientific activities would include drilling activities in order to unlock the history of the lunar surface. The Moon has been bombarded by meteorites over billions of years, and the impacts have thrown up lunar regolith that covers the surrounding terrain. By drilling into the local surface debris, we can reach the underlying unperturbed regolith, which contains a pristine mineralogical record of early lunar events. From this history we would learn about the intensity of the early impacts on the lunar surface, as well as about the lava flows induced by ancient volcanoes. We would learn about the origin of the Moon.

Because of the lower gravity on the Moon than on the Earth and the absence of winds in the atmosphere-free lunar environment, we will be able to build much taller structures on the lunar surface than we can on Earth. There could be super-skyscrapers on the Moon, and not all of them built for commercial activities. Science will have its place.

A novel activity on the lunar surface will be the largely robotic deployment and maintenance of telescope infrastructures built to do science that is unachievable on Earth or even in space. Key steps will involve making use of the vast stable lunar platform and profiting from the lack of any atmosphere to build large-aperture telescopes. These will be telescopes of much greater scope than any that could be constructed on the Earth. Our view of the remote cosmos will achieve unparalleled clarity. We will seek evidence of life signatures on remote exoplanets.

And then there is the radio domain. We need to observe at the low radio frequencies that cannot be discerned on the Earth. We need to go to the far side of the Moon to avoid radio interference from maritime radars, cell phones, and other activities, as well as—even more distracting—the effects of the terrestrial ionosphere, through which the signals would scintillate. These problems are avoided on the Moon, where we would simply see further, closer to our cosmic origins.

Of course, we will need to develop lunar radio-quiet areas given the other lunar activities, including commercial ones. Low-frequency radio astronomy facilities will be among the first lunar science infrastructures to be developed and deployed. They provide the most straightforward of the engineering challenges.

The most remote targets are at such great cosmic distances, and redshifted to correspondingly low radio frequencies, that their radiation cannot easily be studied on Earth. The frequencies are simply too low for high-precision signals. There is too much radio interference at low frequencies. To overcome the noise barriers, radio antennae would be spread over the far side of the Moon to take advantage of the radio-quiet environment and the vast terrain shielded from the Earth. Terrestrial communications would be established by relays of orbital satellites.

The technology is simple. The low-frequency antennae are just two crossed metal bars for a stand-alone system. It looks like a television aerial. For sensitivity, many such dipole antennae must be spread over a large area and connected by radio transmission lines or by laser beams. The large-area deployment provides the high resolving power. Each dipole becomes an element of a giant telescope.

One element in the design that is crucial to constructing an image is to combine coherently the signals from each dipole. Each radio signal is an electromagnetic wave with peaks and troughs. However, they arrive randomly at each dipole, where the incoming wave crests are out of phase. To combine the data into a clear image, we have to combine wave crests by adding them in phase. We can do this with exquisite timing, installing the most precise clocks that are routinely available. These use the vibrations of a crystal to mark time. By allowing us to introduce infinitesimal time delays into the signal at each antenna, such precision enables us to add the signals coherently. We use the principle of wave interference to align the waves. The wave pattern is optimized, so that crests reinforce each other.

The result is a radio interferometer, equivalent to a giant telescope that spans many individual antennae spread over the lunar surface. We can achieve the immense resolving power equivalent to that of a huge telescope diameter. The resulting aperture can be up to hundreds of kilometers across. By adding more and more antennae, we can build a larger and larger telescope. The larger the area, the higher the resolution.

We have successfully designed powerful radio interferometer telescopes on the Earth and use them routinely. The largest telescopes range over several thousands of kilometers, spanning continents. Transferring known technology to the Moon should be straightforward, and deploying large arrays of dipole antennae should not pose an insuperable problem. The technology is simple. The real engineering difficulty is in adding up the signals. A powerful signal processor is needed to combine the data from the many antennae so that we can digitally synthesize the required images.[3]

How futuristic is this? Missions are underway or being planned, implementing small numbers of antennae. The first far-side soft landing occurred in early 2019. Chang'e 4 set up a simple dipole antenna on a lunar rover, having launched Quequiao, a lunar orbiting relay companion satellite, a few months earlier.

One of the goals of Chang'e 4—a collaboration between China and the Netherlands to explore the dark ages—was to carry out low-frequency observations of the radio universe. For technical reasons, little data emerged from the dual-antenna project, but the scene was set for follow-up missions, including projects by both China and NASA. Even the ESA is joining the Moon rush, with plans to develop a lunar heavy-load carrier that will routinely deploy tons of payload on the lunar surface by the late 2020s, for both scientific and commercial use.

One experiment currently undergoing engineering design studies will use a lunar rover to deploy 128 simple dipole antennae on the lunar far side. Principal investigator Jack Burns of the University of Colorado hopes to launch the low-frequency radio interferometer in 2027. With all of these antennae, the local computing requirements to combine the signals are extreme. But given the rate at which computing power is increasing, this goal should be within reach within the next decade. We will supplement lunar surface installations with a flotilla of small satellites orbiting the Moon. These miniaturized satellites, each only a few meters in size, will relay the radio data back to Earth.

A lunar radio interferometer can be designed to provide immense resolving power and is ideal for detecting pointlike sources of radio emission. However, we also need a sensitive radio telescope to pick up the more diffuse signals coming from radio galaxies spread out over the sky. There are many spurious sources of foreground noise in our own galaxy and in similar galaxies. A giant dish is an essential complement to an interferometer if we are to reach out to the dark ages.

To achieve such a goal, the technology is simple. We will build a huge monolithic radio telescope in a crater basin—again, on the lunar far side in order to shield activities from terrestrial radio interference. The telescope surface would be designed by filling in the surface of a large crater basin with wire mesh. Since we will be searching at wavelengths of tens of meters, we will need only the crudest of surfaces. We will suspend the radio receivers in the focal plane by cables spanning the crater rim, which might be five kilometers in diameter or larger. The resulting structure would not be too different from crater-based telescopes we have successfully operated on Earth. But it would be optimized for low frequencies to tune into the dark ages.

Such lunar radio telescopes will enable us to probe the dark ages, the period way back in time before the first stars, just millions of years after the Big Bang, when there were only hydrogen clouds. We will see later that these clouds can be mapped in absorption against the cosmic microwave background, using already known and tried technology. It will not be cheap, of course, but the cost of telescopes is only a small fraction of total lunar expenditures.

Powered by our age-old drive for exploration, and with the dark ages beckoning, we will move to new frontiers. There is no reason why radio telescopes of unprecedented size could not eventually be implemented on the Moon. We have already mastered the technology. It may take perhaps half a century, but the inexorable quest for knowledge will get us there.

Lunar Observatories

Radio telescopes are only part of the story.[4] The Moon offers outstanding sites for building optical and infrared telescopes. Its lack of atmosphere provides spacelike clarity for viewing stars, those brilliant points of light in the night sky. Its low gravity, at 15 percent of that on the Earth, will enable us to engineer enormous telescope structures that will outperform their terrestrial counterparts.

The largest terrestrial telescope has an aperture of 39 meters. The European Extremely Large Telescope is situated on a 3,000-meter mountain peak in the Atacama Desert in northern Chile. This is considered to be the largest filled aperture telescope that can be built on Earth. Plans to build a much larger telescope were abandoned because it soon became apparent that a sufficiently rigid structure could not be easily maintained, owing to conflict between the necessary support structures and the stresses induced by gravity, tremors, and ambient winds. Any structure built significantly larger would be at risk of collapsing.

Such problems are greatly eased by the much lower gravity of the Moon, where larger structures are feasible. In the absence of any winds or significant seismic activity, structural integrity is more easily maintained, and a telescope with a 100- or even 300-meter aperture could be constructed. It would collect 100 times more light than the largest terrestrial telescope and see 10 times as far and with 10 times the sharpness. Construction

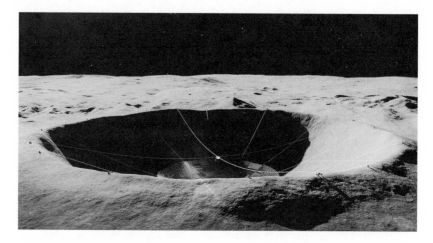

FIGURE 3. The Lunar Crater Radio Telescope would be the largest filled-aperture radio telescope in our solar system; with a proposed diameter of up to five kilometers, it would be much larger than any radio telescope on Earth. It would be built by robots in one of several proposed craters on the far side of the moon. The wire-mesh design will be challenging: it must be strong, flexible, and lightweight, as well as capable of maintaining a precise spacing and parabolic shape while withstanding temperature ranges from −173 degrees Centigrade to 127 degrees Centigrade.

Image credit: Saptarshi Bandyopadhyay/ NASA, https://www.nasa.gov/sites /default/files/thumbnails/image/3_lcrt-crater-view-1041.jpg.

would represent a major step forward not only for astronomy but also, given the discovery potential, for humanity.

The ideal site would be in a dark crater near one of the lunar poles, to take advantage of the permanent cold and continual darkness. Given the polar location, the Sun neither rises much above the high crater rims nor falls very far below the horizon. The temperature is stable throughout the year. The crater rims are almost always illuminated by sunlight, so solar power would be available throughout the lunar year. The crater floor is most likely covered with a thin layer of ice.

We could build the ultimate large-aperture telescope. The telescope structure would be tiled with mirrors, most likely in the form of a paraboloid, with a series of hexagonal mirrors on flexible supports. This structure would simply be a scaled-up version of that of the 39-meter Extremely Large Telescope on Cerro Amazones in Atacama. The ambient low temperature would provide an ideal environment for developing infrared astronomy. Detectors could be easily cooled down further to acquire optimal sensitivity.

But why stop at hundreds of meters? We could imagine supporting the telescope reflectors with the entire crater bowl. If we forgo the conventional design of a monolithic structure, it is the entire span of the dark crater that sets the limiting scale of the telescope. Craters in the lunar polar regions are tens to hundreds of kilometers across. The spacelike conditions on the Moon would help prevent atmospheric degradation of the imaging. We could build a hyperscope with unprecedented resolving power by filling the crater bowl with an array of small telescopes.

One lunar observatory proposal envisages covering the crater floor with a huge network of individual five-meter mirrors. Each would be separately mounted and set in the crater bowl. They would point in coordination at a patch of sky. The signal received would be transmitted to the camera, a focal point detector system suspended by cables spanning the crater rims and mounted kilometers above the crater floor. The images would be combined according to principles similar to the principle of interferometry that underlies combining radio beams. The infrared signal would be focused coherently to produce a single image. It is the required precision of the focusing technology that will be a challenge.

If we succeed in combining the many beams coherently, it will be as though we had a telescope 10 kilometers in diameter.

The great advantage of a hyperscope would be the incredible resolution it could achieve. The resolution would be set by the aperture, which is just the diameter of the crater. We could image the nearest exoplanets in unsurpassed detail, with no need to spend centuries traveling there for a close-up view— not even robotically! One disadvantage of a hyperscope is that it only covers a limited fraction of the crater floor. It's effectively a sparsely filled dish, and we could study only the brightest sources at such high resolution. But the brightest sources would include many dozens of distant planets. We could also look deep into the Universe to image the first stars. Direct imaging would unlock their secrets.

There is a shadow hanging over this vision. For astronomers, it is important that the various space agencies legislate surface and space activities to maintain a dark and radio-quiet lunar sky. The skies above the Earth are already compromised by the proliferation of communications microsatellites. Hopefully, we can do better for the Moon.

CHAPTER 4

Tuning in to Our Origins

The choice is: the Universe ... or nothing.

—H. G. WELLS

At the present rate of progress, it is almost impossible to imagine any technical feat that cannot be achieved, if it can be achieved at all, within the next five hundred years.

—ARTHUR C. CLARKE

In the Beginning

Before the galaxies, before the first stars, there were just vast systems of hydrogen clouds. There were so many back then. If we could detect these systems, our attempts at doing precision cosmology would be greatly boosted. Numbers count. These hydrogen clouds are potential pixels in the radio sky. They are signposts pointing to the past. They will be our probes for studying the dark ages, before the first stars formed.

How did it all begin? No doubt this question arose soon after the first time *Homo sapiens* looked up at the sky. Today our giant telescopes on remote mountain peaks are like time machines that let us travel into the past. To look back in time is to look far away, toward the beginning of the Universe. And so far we have looked using only optical or infrared light.

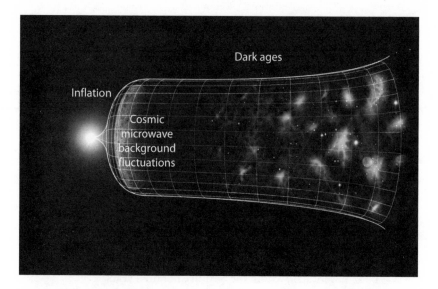

FIGURE 4. From the beginning until today. The time sequence runs from left to right in this sketch of the evolution of the Big Bang. The initial instant, where the physics is unknown, is followed by a rapid phase of inflation that occurred 13.8 billion years ago. From the right, we look back in time to some 380,000 years after the beginning, when we detect the temperature fluctuations in the cosmic microwave background radiation, the fossil glow from the Big Bang. This is as far back in time as we can directly see. The dark ages begin, and as time progresses first clouds form, and then galaxies.

Image credit: Image created by the author.

There is a feeble glow over the entire sky in the form of radio microwaves that come from the beginning of time. Turn on your television and then turn to a channel that is not broadcasting. A tiny fraction of the fuzz on the screen is the microwave glow from the Big Bang. One of the greatest scientific discoveries ever, in 1964, was the detection of the fossil radiation from near the beginning of the Universe.

Even this is not really the beginning, as the radiation gives us a glimpse of the cosmos only after hundreds of thousands of years have elapsed. What was the Universe like previously? We can penetrate deeper into this hazy glow if we are sufficiently clever. How far back can we go? Are there any limits to what profound mysteries might we resolve? The physics of the first moments begins to challenge our understanding of matter itself. Once the Universe was so dense and hot that even particles could not exist. There were no protons, no neutrons. A mere nanosecond after the Big Bang, the Universe was a soup of quarks. Can we access this moment in time?

Finally, we can go on an imaginary journey back in time to explore the beginning of the Universe as it gave birth to all matter, well before the first nanosecond. Probing the beginning of time may be our ultimate attempt to understand nature. As the Universe expanded and cooled, the state of matter changed, like ice melting into water. Just as substantial energy is liberated when ice melts, an energy was injected into the very early Universe that was so powerful and substantial that it inflated the size of the Universe in an instant, to huge proportions. This is the inflation theory, a powerful and paradigm-shifting concept. Let's look at the evidence for it.

All we know about the Universe today points to the beginning as a phase of extreme density. The density was much higher than that of the neutron stars, the most compact stars we have discovered so far. A teaspoonful of neutron star matter weighs trillions of tons. After observing thousands of neutron stars, we have some understanding of physics at this density. Some neutron stars are pulsars, emitting radio blips as they spin. Some pulsars are young, and some are old. Some neutron stars are X-ray stars that accrete gas from close stellar neighbors. As the infalling gas heats up, it glows in X-rays. We can detect many neutron stars with X-ray space telescopes, and X-ray astronomy is beginning to unveil the secrets of neutron star matter.

We have no local examples of the extremely high density at which the cosmos began, a density that exceeded by far the density of the center of a neutron star. Tiny relic black holes may have survived, but they do not tell us much, since any primordial black holes have long since accreted matter from their surroundings and swallowed any clues to information about their origins. The Big Bang theory allows us to track the Universe back to a density that is nearly eighty powers of ten higher than that of a neutron star. That is a density hard to imagine and equally hard to prove.

To understand such an extreme number requires a new theory, one that goes beyond Einstein's theory of gravity. We need to combine gravity, which predicts such a high value, with quantum mechanics. It is the quantum theory that predicts the nature of extreme matter. Only a theory that combines gravity and quantum mechanics can tell us whether such extreme values are ever really achieved in nature.

The problem is that we lack such a theory. We simply do not know how to unify the extremes of density that are achieved as we go from macroscopically large scales, such as the Universe, to microscopically small scales, such as the size of an electron or a quark. Even in the absence of such a unifying theory, however, cosmologists are persistent. We have made some progress.

The quantum theory makes a beautiful prediction. On the smallest scales, uncertainty is paramount. The physics interpretation of the smallest particle size corresponds to the uncertainty in its position. We can think of this uncertainty as a wavelength. Particles are pointlike objects in the macroscopic Universe. Quantum theory views particles as waves. The Planck length corresponds to the scale at which the size of a black hole, the pure product of gravity, is so small that it approaches its own quantum wavelength. This happened very early on in the history of the Big Bang. Indeed, for all intents and purposes, it was the beginning.

We understand the Universe well at an age of one second, and even at one-billionth of a second. We have modeled late-time consequences based on our standard model of physics. We can attain similar energies in particle colliders. Essentially we can try to reproduce the early moments of the Big Bang, but we can never quite approach conditions very near the beginning.

It's much earlier that our extrapolation becomes uncertain and colliders cannot help us. The Planck time—when the visible Universe was just a Planck length across—was a mere 10^{-43} second after the Big Bang. The Universe began at this moment in time. More correctly, that is when our theory of the Universe began.

It is wonderful news that we can approach the moment of creation according to our theory—or more accurately, our lack of one. That's because we lack the ultimate theory that combines gravity and quantum theory. But trillionths of trillionths of trillionths of nanoseconds later, all is well. Our theory is just fine. And what is equally intriguing is that there are testable predictions.

How do we demonstrate that we have access to an instant so close to the beginning of time? Quantum uncertainty is inherent in a tiny particle being a wave. We no longer know the exact location of the particle. This uncertainty leads to infinitesimal fluctuations in the numbers of particles. At any point, particles come and go. Here is a particle—suddenly it has vanished. It is from such tiny fluctuations, quantum fluctuations at the beginning of time, that we have emerged.

These fluctuations of creation eventually seeded the growth of structure. And the coming and going of the fluctuations themselves is a form of dark energy that left a dramatic imprint on the expansion of the Universe. These are the late-time consequences of our theory. But they need to be tested. We will see that the ultimate probes of the fluctuations of creation need lunar telescopes.

The Dark Ages

There was a time when there were no stars. All was darkness. At the beginning of the darkness, the fossil radiation from the Big Bang produced a feeble glow. We can view this today in the cosmic microwave radiation, the highly wavelength-stretched background light modulated by the expansion of the Universe. For every doubling of the size of the Universe in the past, the wavelength of light doubles correspondingly.

The fossil microwave background radiation has propagated directly to us from an epoch about 380,000 years after the Big Bang. We call this the epoch of the last scattering of the radiation. Before then, we view the beginning as though peering through a dense fog. We can see infinitesimal variations on the sky in the intensity of the microwave background. These fluctuations measure the structure of the Universe at this early moment in time.

These fluctuations in light are tracers of the seeds of all structure. They effectively measure energy, and energy, or its equivalent, mass, controls gravity. The associated matter fluctuations are tiny compressions of the density of the Universe. When the matter is compressed, so is the radiation, just as in a sound wave that propagates through space. We cannot directly see the matter, as it is dark, but we infer its presence from the radiation. These infinitesimal density compressions are destined to grow in strength as gravity operates.

At first, the hydrogen is in the form of electrons and protons. The radiation acts like a brake on the ionized gas. But once the density and temperature drop sufficiently, the hydrogen combines into atoms. Atoms are transparent, and the scattering of radiation comes to an end. The atomic era begins, and atoms move freely under the effects of gravity. That is when clouds form, a million years after the beginning. It is a democratic revolution, as the underdogs, the smallest clouds, are the first clouds to form and they dominate. The tiny clouds control the subsequent evolution of structure. We have seen that computer simulations show small clouds joining up with larger clouds in a process of growth from the bottom, that is, a bottom-up process.[1]

This is the period of the dark ages. It takes time, hundreds of millions more years, before galaxies begin to form. But it is inevitable that they will form, gravitationally speaking. The tiny overdensities grow sufficiently overdense to experience the phenomenon of gravitational collapse. The inevitable formation of galaxies and stars demarcates the end of the dark ages. There has been a dramatic transition from quasi-uniformity to the clumpiness of the Universe. Eventually, after a billion years, there are massive clouds, destined to be precursors of galaxies like the Milky Way.

Seeking the Building Blocks

There was a phase before the first stars and the first galaxies formed that lasted for some 10 million years. During that phase there were only cold gas clouds everywhere. These clouds were the building blocks of the future, the mortar to construct the pyramids. The early clouds were ubiquitous. Inevitably they reached the overdensity needed to collapse—first into the dwarf galaxies, then climaxing with the great galaxies of today. But we want to catch them before galaxies formed, at the point when they were the prolific masters of the Universe, the dominant component.

The many clouds consisted mostly of dark matter, which is the glue that holds the clouds together by gravity. Measured throughout the universe, dark matter probably consists of elementary particles that interact so weakly with ordinary matter that they are effectively invisible. One of the greatest challenges in modern astrophysics is identification of these particles. Many different types of searches are underway, thus far with no definitive results. But we are certain that dark matter is ubiquitous, and that it controls gravity.

The dark matter clouds also contain ordinary matter, predominantly hydrogen. About 15 percent of the clouds is hydrogen. Trillions of these clouds were newly condensed from the Big Bang. Gas clouds of hydrogen atoms are embedded in the clouds of dark matter. Atomic hydrogen cools down to about 10,000 degrees Kelvin. That is when gravity holds the clouds together. This limit is determined by the inability of hydrogen atoms to lose any more energy by radiation. It's still too hot for stars to form. But we saw that a small contaminant in the form of hydrogen molecules enables further cooling.

At first, the thermal motions of the molecules resist the gravitational compression of the gas. But the gas inevitably cools. Gravity wins the battle, and the clouds collapse. It takes time, tens of millions of years, for the first stars to form. This phenomenon heralds the end of the dark ages. This is the age of the first light in the universe, the cosmic dawn. It is this moment in time that we hope to survey with the next generation of telescopes. But we can look back even further, where there is no light, into the depths of the dark ages.

Tuning in to the First Clouds

How do we reach back in time to the dark ages?[2] Not with stars—there are none in the first 10 million years. We reach back by seeking their precursors, the first hydrogen gas clouds in the Universe.

These clouds are cold, and as the Universe expands, the gas cools further. Gravity is not effective at resisting the expansion for most of the hydrogen. The dark matter is too diffuse, and the clouds are too small. As such, the cold clouds can be seen as radio shadows against the cosmic microwave background radiation. The shadow is due to absorption by the cold hydrogen atoms of low-frequency photons in the fossil background glow. It is only later that the gas heats up as self-gravity of the dark matter eventually becomes important. That is when the first stars form. Earlier, it is just cold gas. To detect this faint foreground signal, we have to tune in to the Universe at precise radio frequencies as we peer into the dark ages—the first 10 million years.

The radio waves from the dark ages are detected at highly stretched-out wavelengths, that is, at very low radio frequencies. The Universe has since expanded by a large factor, of about fifty. Locally the radio absorption is at a wavelength of 21 centimeters. With the expansion of the Universe, the observed signal has stretched to a wavelength of 10 meters today. Terrestrial or even orbital environments are simply too noisy at these frequencies to detect the tiny signal.

Our proposed lunar radio array would use millions of simple radio antennae, twin-pole structures about 10 meters across. This antenna size is chosen to be comparable to the wavelength of the expected radio waves. It is this match that give the best efficiency of detection, and the observatory will be on the far side of the Moon. Antennae will be deployed and operated by humans in tandem with robots and spread over 100 kilometers of lunar terrain. Data will be transmitted to orbiting satellites to send back to Earth.

At such long wavelengths, the radio signal is extremely faint. The Earth is so noisy with electronic signals that the only way to draw a bead on the beginning is to detect the trillions of data points that correspond to the building block precursors of galaxies. This huge multiplicative factor, or millions of clouds per eventual galaxy, is where high-precision measurements come in. The lunar far side is the only place for a sufficiently powerful low-frequency telescope targeted at the holy grail of cosmology: the dim shadows where there was no light and no stars, but vast numbers of cold gas clouds were everywhere.

Penetrating the Darkness

Exploration of the dark ages is feasible with a sufficiently sensitive radio telescope. The early clouds have atomic temperatures colder than the cosmic microwave background. We observe these clouds at very early epochs, at very high redshift, before any stars, galaxies, or quasars have formed.

We know when to look. If we look any earlier, the clouds are locked into the cosmic microwave background because of scattering. So they maintained the same temperature. The scattering loses efficiency as the density falls while the Universe expands. If we look any later, the clouds are heated as stars form. The atomic hydrogen is soon destroyed by the radiation from the first stars and black holes. All of this energy input raises the cloud temperature to tens of thousands of degrees Kelvin. The hydrogen is ionized again.

So there is a window of opportunity. The first clouds need to be imaged when they were still atomic, at approximately 1,000 degrees Kelvin, and before any stars formed to heat them up. They were briefly in a colder state than the background radiation. This is the optimal moment to trap the desired signal from atomic hydrogen. We can catch the first clouds at about one million years after the beginning of the Universe. If it was not for the absorption, we would never see them. They would emit far too weak a signal to detect in emission.

To actually detect them requires a special approach that permits a probe of hydrogen at very low temperature. Dutch astronomer Hendrik van de Hulst predicted in 1944 that cold hydrogen interstellar atoms should absorb a particular wavelength of radio energy by triggering a slight energy change in the arrangement of the electron orbits in the hydrogen atoms. The energy change corresponds to a precise wavelength of 21.1 centimeters, or a frequency of 1420.4 megahertz. We have since observed this 21-centimeter radiation everywhere in the Milky Way and in many thousands of nearby galaxies. The source is interstellar clouds of cold hydrogen. It is our primary tool for mapping hydrogen throughout the Universe.[3]

We can apply this well-tested technique of radio astronomy to penetrate the dark ages. The idea is to map remote hydrogen clouds against the backdrop of the fossil microwave glow from the early Universe. We seek an absence of light at just this special frequency. The strongest effect occurs long before any stars formed, because that is when the hydrogen clouds were coldest. The 21-centimeter hydrogen line probes the many small neutral hydrogen gas clouds at very early epochs, even before they have collapsed. They are just cold gaseous overdensities. We expect there to be millions of such precursor hydrogen clumps for every massive galaxy. If we could detect them early enough, cosmological precision measurements would be improved by a hundredfold. This represents an enormous step forward in precision cosmology.

A Window of Opportunity

Our challenge is to measure the first clouds in absorption against the cosmic microwave background.[4] Then the wavelength is stretched by the cosmic expansion to something like 10 meters, or a radio frequency of about 30 megahertz. At such low frequencies, it is very difficult to do astronomy. There is too much radio interference under terrestrial conditions. We have argued that the far side of the Moon provides the only solution.

There is more to the story. The vast number of clouds boosts the signal. Cumulatively, we should be able to detect them, given a large-enough radio telescope. But we have to catch them at the optimal moment. If we look too early—that would be less than one million years after the Big Bang—they are too warm. If we look too late, they are almost fully ionized, and there are few remaining cold hydrogen atoms. We already have the first stars, the first black holes, and the first galaxies. The pervasive hydrogen atoms are contaminated by the resulting heat of cosmic evolution. They no longer reliably trace that part of the past since 100 million years post–Big Bang.

So we have a natural gap in which to look into the past: between one million and 100 million years after the beginning. This translates into a window of opportunity in terms of redshift, another measure of time. Redshift is the lengthening of the wavelength of light to the red that measures the cosmic expansion of space. There is a sweet spot, when the shadows are deepest. We need to detect the hydrogen at a redshift factor of around 50. The Universe was then fifty times smaller. The 21-centimeter wavelength of hydrogen atoms is stretched to an observable wavelength of about 10 meters, or a radio frequency of only 30 megahertz.

Thirty megahertz is a really low frequency for radio waves to be able to cleanly traverse the Earth's atmosphere. It is the ideal frequency for an absorption signal, however, because the clouds are then at their coldest compared to the background light. We can search for their imprint. In effect, we are listening in to the Universe, seeking to extract its deepest structure. Such a low frequency can barely be observed on the Earth, and then only with the greatest difficulty. There is too much radio noise, both natural and man-made. The far side of the Moon is the ideal listening post.

Normally we can justify investment in a new particle collider or new telescope if we can increase sensitivity by a factor of 10. The preferred ballpark criterion for a major investment is an increase by a factor of 100. We expect to achieve this advance with our lunar strategy. We will do cosmology on the Moon and study the structure of distant clouds with unprecedented accuracy that will allow us to unlock the secrets of the beginning.

First Steps

A major effort is currently underway to study the end of the dark ages, using terrestrial telescopes, as the radio frequencies are within reach. This is the simplest way to detect the transition between the cold, dark universe and the onset of cosmic dawn, when the first stars formed. Even here, it rapidly gets complicated as more and more pockets form of heated and ionized hydrogen. Soon all the intergalactic gas is ionized. Few atoms remain.

The easiest way to see this transition is by searching for an edge in the redshifted hydrogen signal. First the cold gas is there, but soon it is gone, transformed into ionized gas. We will not have enough antennae to see individual clouds, but even with a single antenna unable to give any detail, we should be able to see a steplike transition as the cold gas vanishes with the formation of the first stars. This is the instant of cosmic dawn.

We can do rudimentary cosmology with a single antenna, and in fact several such experiments are underway. A special terrestrial environment is needed, typically that of a remote island. It must be free from human sources of radio noise. Another obstacle is the scattering of radio emission from the galaxy into the antenna beam as radio waves bounce off the ground. With only one antenna, it is hard to control the scattering from distant sources. The surrounding water give much better control of spurious radio wave reflections that can corrupt the elusive signal.

Another approach is to build a compact network of antennae, with the dipoles almost touching each other. The antenna signals are combined, and the number of elements ensures sensitivity and resolution. The more antennae, the higher the resolution. Such systems are under construction in South Africa and Australia.

HERA is being built in the Karoo Desert of South Africa, where radio signals from the indigenous population is minimal. Targeted will be radio frequencies of around 100 megahertz, which corresponds to the epoch of the redshifting of the hydrogen 21-centimeter line when cosmic dawn begins. From the Earth, it is difficult to go earlier, which requires observing at lower radio frequencies. Up to 400 dipoles will be incorporated, but this is far from enough to resolve the cold gas cosmic building blocks. Rather, the intent is to measure the onset of cosmic dawn. The signal strength will indirectly tell us about the nature of the transition between an atomic and an ionized universe. But it will be a blurry picture.

The lunar far side provides the ultimate environment in the next decade for detecting a global signal of cosmic dawn. Radio contamination can be much better controlled there than in even remote parts of the Earth. Hundreds of thousands of antennae can be assembled on the floor of a large lunar basin on the far side to build a radio telescope of unprecedented resolution. Above all, it will have the low radio frequency range that is needed to dive into the dark ages but that cannot effectively be accessed from the Earth. We will then see that our lunar radio telescopes have the power to unlock the most profound secrets of cosmology.

Keeping Perspective

Let's stop here for a second. When I began to work in the field of cosmology about a half-century ago, we were discussing precisions at best of 50 percent for any measurable parameter. I wrote my doctoral thesis on applications of the newly discovered cosmic microwave background to galaxy formation theory. The data were crude. Cosmologists did not know the expansion rate of the Universe to better than a factor of two. They did not know the dark matter content of the Universe to any better than a factor of ten. We were living in the dark, cosmologically speaking.

On the positive side, we were asking the right questions. But because we were very far from precise answers, skeptics abounded. One needed a certain degree of faith in unproven theories to accept the Big Bang. And even more faith to accept that it was initially so hot.

Cosmology has changed dramatically in the past decades, especially thanks to cosmology telescopes on space satellites. These range from COBE, launched in 1989, to WMAP in 2001 and PLANCK in 2009.

The perfect blackbody nature of the fossil microwave background pointed unequivocally to a hot origin. Long ago, in my doctoral thesis, I predicted that there must be fluctuations. Otherwise, we would not be here. No galaxies would have formed. Over several decades, new generations of experiments mapped the fluctuations in the microwave sky. We emerged into the sunlight, having finally found direct evidence for the Big Bang. There are always skeptics, but they could now safely be ignored. Experiment trumped fantasy. Precision approaching 1 percent became the norm. It is an amazing advance that cosmology has become a precision science.

Fossil Glows

Here is how future searches will be optimized to test inflation theories. Next-generation microwave background experiments are targeting a primordial twist in the signal from the epoch of inflation. We call this signal polarization, and it is our best theory of the beginning. It can be seen in scattered sunlight, where polarization is induced by dust. Change the polarization angle with a special filter of polarized glass and the glare is removed. This polarized glare is a property of the dust, and it is what allows us to detect the dust.

Polarization of the microwave background is due to ancient gravity waves produced in the first instants of the Universe. The signal is imprinted onto the fossil radiation when the Universe was last dominated by radiation. We see this signal in the difference with a pure blackbody spectrum that is unpolarized. The passage of gravity waves preferentially stretches the microwaves in one direction and imparts a polarization. We can detect this imprint in the distribution of fossil photons. Detection would verify a key prediction of inflation cosmology.

The cosmic microwave background temperature fluctuations amount to a few millionths of a degree Kelvin. That is a small fraction of the 3 degrees Kelvin we measure for the relic microwave background. Painstaking efforts by my experimental colleagues led to this triumph of detection and mapping. But we can do better. Ultimately, the most accurate measurement of the cosmic microwave background is just one divided by the square root of the number of pixels. With this measurement, there is no more information to be obtained. This limits our precision for any measurement designed to extract the key parameters of cosmology to at best one part in 1,000. That is still ten times more precision than the Planck satellite achieved.

We will achieve this ultimate precision in the next generation of microwave background experiments, which will be performed in the Atacama Desert, at the South Pole, and on space satellites. Yet the data resource is limited. Ultimately, the signal comes from about three million independent points on the sky. We uncovered the seeds of creation, the fluctuations that seeded galaxies, and are striving to do better. But we risk running out of information in the microwave sky.

The predicted polarization signal takes us well into the darkness. The slight twists actually probe a much, much earlier phase, well before the last scatterings occurred. At least they do so in some theories. One interesting conclusion is that there are many theories that fail to have the elusive twist. That is a reminder, of course, that we have to look even deeper. But there is no guarantee of success.

We need to do better in order to verify the predictions of our leading theory for the Big Bang. Tracking the elusive microwave polarization signal is not enough to tackle questions of our origins with any high degree of confidence. It is not a guaranteed outcome of the inflation theory. The next decade will give us all the precision we need to confirm our basic model of the Universe. This will be a wonderful achievement. Much more accuracy will be needed, however, to test the greatest question of all—the cosmic origin hypothesis.

Did we begin in a Big Bang? By far our best contender is indeed the inflation theory. To test this we will need to look much more deeply and sensitively into the past. Suppose we fail to detect the elusive polarization signal. A real concern is that there is no guaranteed signal. If there is no detection, what next? We need to do better, to collect more information. We must consider other strategies.

Counting Galaxies

Let's consider counting galaxies.[5] Cosmic microwave background studies of fluctuations focus on the evolution of the Universe before there were galaxies. Galaxy surveys zero in on the nature of galaxies, their environment, and their evolution. As the data quality improves, we begin to see the initial fluctuations in density from which large-scale structure grew. We need both approaches to the primordial fluctuations. The science goals are highly complementary. Galaxy surveys have more information content than the microwave background radiation fluctuations. There are many billions of galaxies. Let's see what counting galaxies can do for us.

Planned surveys with a new generation of telescopes will identify up to several billion galaxies. This is a big increase in information content over previous data. The plan is to use megacameras with billions of pixels. We are on the verge of a revolution in galaxy surveys with the next generation of telescopes, both on the ground and in space.

Each galaxy is different. We need a minimum number, at least 100 galaxies, to have a reasonably independent sample of information bits of galaxies. So I estimate that there are about 100 million independent samples. This is a huge gain over the information content in the microwave sky. Galaxy surveys increase the accuracy with which we can now attack the origins of the Universe by as much as a factor of ten, compared to the cosmic microwave background.[6]

Again, it is always the square root of the number of data points that controls the uncertainties and identifies the uncertainties in any sample of independent bits of data. This is why we need large samples in a population census or in a medical trial. Numbers are crucial for accuracy. In the realm of the galaxies, there are many effective pixels, now 100 million in the sky, with the related precision. With the new telescope facilities, we will take the next big step forward in cosmology. The billions of galaxies in the forthcoming surveys will take us back to a detailed picture of the Universe at half its present age.

But this still will not be enough to take us where we would like to go, to probe the nature of the beginning, because we simply run out of galaxies that can be imaged with the needed precision. We run out of information pixels with any galaxy signal. We need even more data. The real challenge lies further ahead, with the great telescopes of the future that are yet to be designed. Even their data will be limited in information content. We will still run out of galaxies.[7]

We will see next that only hydrogen clouds in the dark ages give a guaranteed signal that meets the ultimate challenge of inflation. The dark ages will give us the most powerful probe of the beginning. Lunar telescopes will have an important role to play in this probe.

Mapping the Beginning

We must extract all the possible information that there is from the elusive signals. There is really only one direction left to explore: the opening up of the dark ages. That is where so much more information is lurking. We have seen that the far side of the Moon offers a location to achieve the best we can ever do. To achieve our goal we will use the maximum number of independent bits available.

Bits are the key to sensitivity. The number of independent bits is a measure of information content. Counting bits is equivalent to counting the number of pixels in a CCD camera image. The image contains all of the information. Higher resolution reveals more information and requires more pixels. The more pixels in the digital camera, the higher the resolution that is attainable. In the drive for the ultimate sensitivity in a telescope, it is resolution that counts above all. We want to collect photons from as many targets as possible, even though ultimately we are limited by the number of detectable galaxies.

There is only one way to beef up the precision of cosmology: get more information. With millions of clouds per typical galaxy, the dark ages are our only hope. They are completely unexplored territory. They certainly present an enormous challenge, but they also offer a unique glimpse of the ultimate probe of the beginning.

The ultimate precision will come when we optimize the information content of our signal. Probing the dark ages will open up a trillion bits of information and allow a huge increase in precision over surveys of all the galaxies in the visible universe. Typical future surveys will target billions of galaxies. By moving to the dark ages and focusing on low-frequency radio signals, we can take a giant step forward.

We simply cannot afford to miss this opportunity, which presents the most serious challenge ahead of us. The 21-centimeter line of atomic hydrogen is a clean probe of the dark ages. We will measure the distribution of cold gas, the hydrogen destined to form the first galaxies. We will catch the gas in the cosmic womb, just prior to its utilization in the building blocks of the galaxies. The dark ages provide a unique glimpse of cosmic history because there are no stars or other sources to contaminate and heat up the primordial hydrogen. And because the numbers of these primordial clouds are extraordinarily large—millions per typical galaxy—information content will be hugely enhanced over any galaxy survey.

Terrestrial telescopes, even one as large as radio astronomy's Square Kilometer Array, are plagued by man-made interference at low frequencies. The earth's ionosphere is another major obstacle at these cosmologically obligatory low radio frequencies. I have argued that the far side of the Moon offers a unique environment for a low-frequency radio array and that lunar radio telescopes are our best option for realistically exploring our beginnings. What is the best use of this new precision that awaits us?

Beyond Bell Curves

Our leading theory tells us that the Universe briefly inflated long ago, imprinting a tiny and unique distortion on the distribution of the hydrogen clouds.[8] In any consideration of this theory, it is all a question of not being random. True randomness is actually rare unless we go to really large samples of data. Many processes in nature are intrinsically nonrandom. It is when these processes combine that randomness emerges. Randomness hides the intrinsic secrets of nature. We will need to unlock these secrets to come to grips with the leading theory of the beginning of the Universe.

The mathematical term for not being random is non-Gaussian. The German mathematician and astronomer Carl Friedrich Gauss, who lived from 1777 to 1855, invented the bell curve, which is a measure of randomness. The bell curve refers to the distributions of independent samples of measurements or just of some random variable. For many independent observations or data sets, the spread in the variable being measured always converges to a symmetrical curve. It does not matter what we are measuring, just that we measure it many times. The bell curve is an incredibly useful predictor in statistics, the social sciences, and the natural sciences. It is used in demography, epidemiology, psychology, and economics. Likewise, the bell curve is applied in physics and astronomy as relic signals from the beginning are examined. Like the old adage that all roads lead to Rome, including enough channels in any process leads to a bell-shaped distribution.

Another of Gauss's insights that he developed so far ahead of his time is that we can mathematically transform any complex image into a set of numbers. Numbers can capture the full information content in an image. The Mona Lisa portrait is highly non-Gaussian. It can be degraded to a large set of numbers that are very much nonrandom. The numbers capture the beauty and subtlety of the painting. If we randomly mixed up this set of numbers, we would recover something resembling a Jackson Pollock painting, at best. Likewise, we would transform the digital representation of a Bach sonata by randomization into a piece more closely resembling something by, say, John Cage. Of course, even these examples still would not be random. Far from it. Let's not deny the artistic genius of a Pollock or a Cage.

Pure randomness is pure noise. We may infer that essentially all of the initial subtleties—indeed, all of the initial beauty—lies in the non-Gaussian aspects of a piece of art. It is not easy, however, to compare an artwork with truly random noise. Pure randomness is incredibly hard to design by any artificial process. Of course, any piece of modern art or modern music is far from a Gaussian distribution of patterns. There are many non-random bits to the information they contain. But the patterns and symmetries are less dominant. The harmony of a Verdi opera, a Beethoven symphony, or a Rembrandt portrait is certainly lacking. It's a question of the information's phases or symmetries, which are very nonrandom and contain aspects that are far more dominant than in the case of abstract art. Or at least, that is my subjective impression.

The Holy Grail

Now let's turn to the very early Universe. Our models generally have a distinguishing element of non-Gaussianity that can provide a genuine test of the earliest instants. But the predictions are completely masked by highly non-Gaussian foregrounds. Consider radio waves as a possible probe of the dark ages. The foregrounds are determined by the earth's atmosphere, ionosphere, maritime radar, cell phones, TV broadcasts, radio transmissions, microwave ovens—and the list continues. All of these relatively local sources of noise mask any background signal. Further afield, there are the radio emissions from Jupiter, the Sun, the Milky Way, and the galaxies far beyond. We need to dig deep into this foreground noise.

Our Universe is highly nonlinear, in the sense that galaxies are collapsed objects. The formation process itself generated non-Gaussianity. Some regions collapsed, others did not. Galaxies often formed in clusters and groups. On the larger scales, galaxies are found along a network of large-scale filaments that spread throughout the observable Universe. Galaxies are nonrandomly distributed in space. Almost all galaxies emit radio waves, resulting in a pervasive glow of radio emission everywhere there are galaxies. And their large-scale alignments and groupings make the glow intrinsically non-Gaussian. The galaxies contribute a foreground that obscures the distant signal from the early Universe. It is like driving a car with a dirty windshield. We have to eliminate all of these traces to look deeper into our surroundings and into our past.

Astronomers cope with foregrounds by superimposing maps of galaxies onto the measured signal. We digitally clean the images of foreground distractions by looking for spatial differences in the distributions. Awaiting us as we follow this procedure is the underlying holy grail, the primordial non-Gaussianity, the only robust and undisputed prediction of the inflation of the early Universe. All models predict this, some more than others. The future goal is to search for this elusive signal. Only low-frequency radio telescopes on the far side of the Moon will provide enough raw sensitivity.

We seek the elusive signal using all possible tracers of structure to clean our maps. Noise reduction is the key. We achieved miracles with the cosmic microwave background. The signals at the discovery epoch were in degrees Kelvin. The Planck telescope achieved ten-micro-Kelvin sensitivity to fluctuations in the signal. The sensitivity improved nearly a millionfold. And forthcoming developments will increase the sensitivity a thousandfold. But this will not be enough to robustly test inflation.

Levels are needed for the diffuse low-frequency radio signal that are 1,000 times more precise than ever attainable with the cosmic microwave background. We are information-starved. We are far from the needed sensitivity in any foreseeable galaxy catalog. There is not enough information. The dark ages' sensitivity would guarantee the only sure signal from the beginning of the Universe. Only failure to find such evidence can falsify our leading theory.

There is reason for optimism. Imagine looking at the stars near a streetlight. The dazzle is more than distracting—it is blinding, and we can see nothing fainter than the stars. Similarly, overcoming the radio dazzle will be our major challenge. But we have solved similar challenges before. Ultimately it will be lunar-based technology that allows us to peer into the dark, to see how close we can get to creation.

CHAPTER 5

The First Months of Creation

The evolution of the world can be compared to a display of
fireworks that has just ended: some few red wisps, ashes and
smoke. Standing on a well-chilled cinder, we see the slow fading
of the suns, and we try to recall the vanished brilliance of the
origin of worlds.

—GEORGES LEMAITRE

How It All Began

Our greatest breakthrough in understanding the beginning of
the Universe came with the theory of inflation. It was driven by
the need to better understand particle physics in the cosmic
laboratory of the very early Universe.

There are four fundamental interactions in physics: the electromagnetic force, the weak nuclear force, the strong nuclear
force, and gravity. The earlier we go in time, the hotter is the
Universe, and the higher the energies of particles. At some sufficiently early point in time, above a really high energy, the
forces are unified. We are striving to understand how this works.

One force combines electromagnetic and nuclear interactions at what we call the grand unification scale. It happens at
an energy equivalent to a quadrillion proton masses. At higher
energies, these forces are indistinguishable. This is physics that

relies on simple extrapolations from the low-energy world with which we are familiar. We can do even better. Even earlier in time, at a thousand times higher energy, gravity joins the party. This is the ultimate union of the forces. The only problem is that we do not yet have a unifying theory that includes gravity.

Concordance, elusive though it is, should set in at a certain instant in time that is just millionths of a trillionth of a trillionth of a trillionth of a second after the beginning. That instant is the effective beginning of the Universe. As Georges Lemaitre put it so elegantly, there was no day before yesterday. We have a name for this moment—the Planck time—but we do not really have a clear view of the physics. Before then, our theory breaks down.

We would love to believe that gravity is unified with the other forces above the Planck energy scale. Above this energy, however, we are in the murky realm of quantum gravity, where gravity and quantum theory merge. Of course, that has never stopped cosmologists from speculating.

Inflation occurs when the Universe is teeming with the effervescence of particles and antiparticles that come and go over immeasurably small times. This creates a sort of quantum foam in what otherwise is the vacuum. In other words, the vacuum has energy. This phase does not last long; the quantum fluctuations are over as soon as space expands and the matter cools slightly. But an unforgettable trace is left behind.

The energy of the vacuum is a manifestation of what Einstein called the cosmological constant. Einstein and Lemaitre realized that this constant acts like antigravity as it counters the decrease in matter density ordinarily produced by the expansion of the Universe. Before observers persuaded him otherwise, Einstein favored a balanced, non-expanding, static universe. Lemaitre went all the way to endorse acceleration, thanks to the predominance of the cosmological constant. He realized that it manifests itself as a form of energy density of the vacuum that is a form of dark energy, thanks to quantum fluctuations. And most importantly, he realized that the energy of the vacuum acts like a negative pressure. If large enough, it turns out, Einstein's theory shows that negative pressure accelerates space. And the negative pressure was indeed large enough to compensate for and eventually dominate over the gravitationally attractive effect of ordinary matter and does just this at late times.

The acceleration of the Universe was largely treated as a curiosity before reinforcements arrived. A notable supporter was the eminent astronomer Sir Arthur Eddington. However, the late domination of the cosmological constant was a minority view. It was to take half a century before new insights into acceleration had a revolutionary effect on theoretical cosmology. And it would be another two decades before Lemaitre's ideas on the meaning of the cosmological constant as dark energy were vindicated by observational evidence.

The next step required new players in cosmology from another field of physics. Just as in the 1950s nuclear physics had a huge impact on our understanding of the evolution of the chemical elements, it was now the turn of particle physics to change forever our understanding of the beginning of the Universe. The theory of cosmic inflation, born in 1981, was in essence the acceleration of space by vacuum energy. Inflation lasted only for the briefest instant of time as the quantum era ended. But its imprint has persisted ever since.[1]

The fleeting moment of acceleration is enough to illuminate space, via the propagation of light and hence information, to a vastly larger scale than the initial value of the actual horizon. The acceleration is rampant. Space expands at a speed much faster than light speed for the briefest of instants. The vistas of the distant Universe open up. As it expands so dramatically, space homogenizes, like the smoothing of wrinkles on an inflating balloon. But the tiniest ripples remain. The quantum fluctuations are stretched from infinitesimal to much larger scales as space inflates. There are virtually no limits on the size of the visible Universe; huge vistas of space open up. Very soon, however, the curtain comes down as the rapid expansion phase of inflation ends. The normal expansion rate resumes in a Universe no larger than the light crossing time since the beginning, limited to just the tiny fraction of a second elapsed.

This helps explain why the Universe far away looks much like the local Universe. Space carries the imprint of its inflationary beginning. We can understand why the Universe from our present perspective looks the same in all directions. Distant regions communicate by gravity. We can account for the immense size of the Universe; long ago it became huge, albeit for the briefest instant of time. But that's all that's needed. Inflation explains why space is indistinguishably close to being flat or Euclidean as a consequence of the enormous expansion. Any large wrinkles in the curvature of space disappear.

All the density fluctuations that seeded structure had a quantum origin. The origin of the fluctuations is an otherwise impenetrable mystery of the beginning. Previously, cosmologists blamed density fluctuations on initial conditions. They are on scales that once were totally out of communication in the first instants of the Universe. Thanks to inflation, tiny fluctuations in density are stretched onto enormous scales. Only at much later times do these reenter the observable Universe, renew contact with each other, thanks to gravity, and grow into the galaxies.

On average, the fluctuations everywhere are similar, they are statistically indistinguishable, and they look identical. This is an experimental fact. The fluctuations have been detected and mapped as millions of tiny temperature ripples in the cosmic microwave background radiation. These blemishes on the microwave sky are the seeds of all large-scale structure we observe in the Universe. But we will see that when we run out of microwave sky to explore, only the Moon may provide access to the next frontier.

Confronting Reality

These are the miracles of the inflationary beginning. This is great news for theorists. There are now answers to some of the most fundamental questions about our origins. But how can we actually prove any of this? It all happened so long ago.

Our best bet so far has been to study the tiny temperature fluctuations in the sky. Their distribution of strengths is almost the same on different scales, except that there is a tiny systematic increase toward the short-scale end of the spectrum. That is what we infer from the data on the cosmic microwave background. Both the existence of galaxies and inflation tells us that there must indeed be deviations from homogeneity. Those are the fluctuations. In the simplest theory, these should be extremely small. Indeed, it took decades of painstaking effort to detect them.

It is the observations that set the scene on large scales. Inflation is just a good phenomenological fit, with very few parameters, to the data. It takes just six numbers to match the observed Universe. This fit has been one of the great success stories of modern cosmology. But it is still not good enough. On small scales where there are no data, the fluctuations could have any value. They are completely unconstrained by observations of the fluctuations in the cosmic microwave background. So the data demonstrate the principle of inflation but do not do much more.

The fossil radiation approach limits precision in determining cosmological numbers to at best a percent. Of course, this is a remarkable advance since the discovery of the cosmic microwave background. Yet it falls well short of what is needed in order to search for the fingerprints of inflation. We would like to probe the very beginning of the Universe. The only guaranteed signal requires a hundredfold or even a thousandfold increase in precision. This will not be easy.

The end of inflation is where the cosmic journey begins. We have few probes. Indeed, there may be only one robust prediction. To make progress we need to probe the final unexplored frontier of the early Universe, the dark ages. This epoch provides our clearest glimpse of the Universe before any stars formed. The gaseous building blocks of galaxies were all that existed in terms of structure. These are viewed as fossil shadows against the primordial radio glow from the Big Bang. They are ghosts from the past. The expansion of space lowers the frequency of these radio waves to the limits of what is observable. Studying these gas clouds at such early times can only be done at very low radio frequencies. We will see that this is achievable via telescopes on the lunar far side. This will lead us to the ultimate probe of inflation.

The fluctuations in the hydrogen absorption signal are not totally random. They have some slight asymmetry in their distribution of strengths. The asymmetry amounts to a primordial deviation from the usual bell-shaped curve. That is, inflation predicts tiny deviations from randomness in the initial density fluctuations. This effect has yet to be measured. But it is a solid prediction, true for all inflation models. It is a very small effect and requires a huge increase in our current experimental sensitivity. It is the ultimate test of inflation. A telescope on the far side of the Moon is by far our best bet for achieving the goal of detecting deviations from primordial randomness on different scales.

We cannot be confident that inflation occurred. Inflation remains a conjecture about what happened shortly after the Big Bang, albeit one that most physicists believe. One of our principal aims is to show how the next stage in human exploration, that of our nearest neighbor in space, will provide a unique opportunity to probe our cosmic beginnings.

Before the Beginning

Here is a slight diversion before we grapple with the practical aspects. A question that everybody poses when confronted with the scientific theory of the creation of the Universe is this: What happened before the Big Bang? There are two approaches, one based on religion, one on science. Both are pertinent because the founder of modern physical cosmology was motivated by these seemingly contradictory pathways. A century ago, Georges Lemaitre managed to reconcile the two orthogonal aspects of cosmology. He went on to predict what is now the standard model of cosmology—an expanding and accelerating universe.

First, however, let us go back even further, as far back as the fifth century AD, for an answer that still rings true today. The question of a cosmic origin was posed by Aurelius Augustinus, Bishop of Hippo, in what now is Algeria. He died in AD 430 and was canonized in 1298. Augustine developed into a remarkable philosopher and theologian, and his contemplations on many deep issues about the nature of the Universe have survived in his writings. Of special relevance to the creation of the Universe, he posed the question: What was there before?

Saint Augustine did not respond, as often misquoted: "God was preparing Hell for people who asked such questions." More pertinently, he actually wrote:

> There can be no doubt that the world was not created in time but with time. An event in time happens after one time and before another, after the past and before the future. But at the time of creation there could have been no past, because there was nothing created to provide the change and movement which is the condition of time.

Fast-forward to modern twentieth-century cosmology. Views about the earliest instants of the universe are succinctly expressed by one of the founding fathers of the Big Bang, Monsignor Georges Lemaitre. After completing his doctoral studies in general relativity, Lemaitre was ordained in Belgium as a priest. Interestingly, as an ordained priest, he reconciled the Big Bang with his religious beliefs by arguing that "it appeared to me that there were two paths to truth, and I decided to follow both of them."

Most modern theorists argue that, indeed, time and space began in the first instants of the Big Bang. But there is still no consensus on how this was actually achieved.

And the Universe Expanded

In 1927, Lemaitre published his theory of the expanding Universe. He demonstrated that it was favored by data on redshifts and the distances of nearby galaxies. He used the data to demonstrate that the Universe was expanding. This theory implied that there was a beginning. Lemaitre later became a cardinal and a science adviser to Pope Pius XII, who published the 1950 papal encyclical that welcomed the new insights into creation offered by science. The historical records are incomplete, but it is thought that Lemaitre urged caution in overinterpreting such connections.

Previously, accepted dogma held that the Universe was created about 5,000 years ago. One biblically well-researched case was made for October 24, in 4004 BC, by Bishop James Ussher in his *Annals of the World*, published in 1650. Today's creationist movement continues to support this biblical interpretation. However, the major religions of the world now accept the expanding universe, and its corresponding age of some 14 billion years, as fact.

Of course, the revision of the age of the Universe was not a huge surprise. Knowledge of the long geological ages of the Earth were widespread in the nineteenth century. Early in the twentieth century, radioactive dating of uranium-rich rocks confirmed a long timescale of at least several billions of years. The argument from the cosmic perspective resonated well with the earlier debate. A new paradigm for cosmology was signaled.

Along with solutions of Einstein's equations performed independently by the Russian scientist Alexander Friedmann in 1923, Lemaitre's early work marked a turning point in our understanding of the Universe. The new expanding cosmology required a very old age for the Universe. That was readily accepted in view of the geological evidence for an old Earth. Lemaitre eventually became famous as the co-discoverer, with Friedmann, of the expanding Universe cosmology.

American astronomer Edwin Hubble, unaware of the young Belgian scientist's pioneering work, would re-derive Lemaitre's result on the galaxy-redshift distance relation in 1929, motivated by new observational data on galaxy distances. A year later, Lemaitre's former supervisor, Sir Arthur Eddington, had

Lemaitre's paper translated from French. Only then was the new cosmology widely popularized. Einstein, an earlier skeptic about the expanding Universe—or more precisely, the expansion of space—was convinced. He is said to have commented after hearing Lemaitre's account in a lecture in Pasadena in 1932: "This is the most beautiful and satisfactory explanation of creation to which I have ever listened."

As evidence of the expansion of the world of galaxies accumulated, the Big Bang theory gradually supplanted the earlier ideas about cosmology. The final step in the validation of Lemaitre's reasoning came with the discovery, half a century or more later, that at late times the Universe is indeed accelerating. Dark energy dominates the expansion, albeit only during the last few billion years. And the data, improving in precision over the past two decades, converge on the cosmological constant being just that—a constant addition to the energy density, and one so small that it only very recently in cosmic history came into its own as the Universe expanded to its present low density. Surely Lemaitre would have been delighted, even if the dark energy density was very different from what he had envisaged.

We can date the discovery of the acceleration of the Universe to 1998. Two competing groups of astronomers independently found evidence for the cosmological constant. Both found that distant supernovae were about 20 percent fainter than they should have been. Supernovae, or exploding stars, of a certain type were believed to be incredibly precise beacons, exploding via the decay of half a solar mass or so of radioactive nickel created in the precursor star's collapse. Because of their immense luminosity, such supernovae are visible far away in distant galaxies. The systematic dimming of the otherwise identical distant supernovae could only be explained by the acceleration of the Universe. The astronomers demonstrated conclusively that dimming by distant dust was not the culprit. There was indeed a nonzero value of the cosmological constant. For this discovery, the Nobel Prize in Physics was awarded in 2011 to Saul Perlmutter, Brian Schmidt, and Adam Riess.

Over the intervening decades of modern cosmology, pioneered by Lemaitre and Hubble, our understanding of elementary particle physics has advanced considerably. Our best attempts to calculate the value of the energy of the vacuum have exposed a glaring gap between prediction and observation. That difference indeed represents what some of my colleagues have dubbed one of the greatest puzzles in modern physics. Why is the cosmological constant so small? And why does it only begin to have an impact in the recent past?[2]

There are recent hints that new but unknown physics might be needed. There is a disagreement among astronomers about the measured expansion rate. Differing techniques disagree at a level of about 10 percent. And the discordant results argue for measurement uncertainties much less than this difference. To add spice to the discordance, one set of data emphasizes the early Universe, via the cosmic microwave background radiation, whereas another focuses on measurements of nearby galaxies.

This might be a pointer to new early Universe physics. I personally believe that the data issues are far from being robust enough to motivate new physics. For me, the cosmological constant is just one more constant of nature, along with others that have yet to be understood because we do not yet have a theory. And any observational tensions are likely to be due to our failure to account properly for observational biases.

There are certainly issues with our best theory of cosmology, the Big Bang. The huge densities attained near the beginning signal the breakdown of classical physics. We need to merge gravity with quantum theory, but the final theory that unites the quantum theory with gravitation is still beyond our reach. It is the focus of intense research efforts. Presumably, that theory will someday tell us the value of the cosmological constant, if it is indeed a constant.

So far, attempts to derive the ultimate theory—often referred to as the "theory of everything"—have been unsuccessful, in large part because it is really difficult to devise experimental tests of such a theory. The energies attained at the beginning of the Universe far exceed anything attainable in the largest conceivable particle collider. Clever ideas are needed. We cannot probe the beginning, but we can still do much to explore what happened just after the beginning. Here is where lunar telescopes again rise to the fore.

Probing the Celestial Fireball

The most perfect radiation ever created is in the sky. We can think of it as being generated in a gigantic furnace during the first months of the Universe, when the Universe was at millions of degrees Kelvin. Today we measure the temperature of the Universe at about 3 degrees Kelvin. The radiation cooled down as space expanded. The Universe contains a mixture of matter and radiation. In early times the Universe was dense and hot. We are immersed in the cold relic of the primordial fireball. Once it was a perfect furnace filled with blackbody radiation.

How do we know this? The 3-degree-Kelvin radiation is most intense at microwave frequencies. Radio astronomers measure the radiation from the sky, then subtract all extraneous signals caused by the Milky Way and other galaxies. What remains, less than a percent, is uniform across the sky. The distribution of frequencies of the background radiation is that of an ideal blackbody. Known sources of radiation, such as dust emission, add little to the cosmic radiation budget. The cosmic background glow is so precise a blackbody that it could only have been created in the dense and hot environment of the very early Universe, within weeks or months after the Big Bang. Only then was the Universe dense and hot enough to guarantee that all possible energy sources resulted in a perfect blackbody.

The coldest place in the solar system was long thought to be the dwarf planet Pluto, the outermost planet from the Sun. But Pluto's chill has been overtaken by our nearest celestial neighbor, the Moon. There are cold traps in permanently shadowed craters near the lunar poles that are really cold. This is where we would like to put our future lunar telescopes. At low temperatures, detectors are especially sensitive to infrared radiation.

The Universe is now a cold place. Once it was very hot, a long time ago. Soviet physicist Yakov Borisovich Zeldovich was one of the founders of hot Big Bang cosmology. He was a prolific contributor to theory around the time of the discovery of the cosmic microwave background radiation. Zeldovich was impressed by the difference between the 3 degrees Kelvin at which we measure the cosmic microwave background and its inferred temperature in the past, when the Universe was much denser. In his own words, "The point of view of a sinner is that the church promises him hell in the future, but cosmology proves that the glowing hell was in the past." Perhaps Zeldovich wrote this in part as a reflection on his early career as a successful designer of the first Soviet hydrogen bomb.

Why are we so confident in this immense extrapolation back in time, from our perspective of 14 billion years later? So much time has elapsed, yet we have overwhelming circumstantial evidence, coming from the cosmic abundances of the lightest elements, most notably the second most abundant element in the

Universe, helium. It amounts to 30 percent of the mass of hydrogen.

The incredibly high temperatures in the Big Bang lasted only for minutes or hours. This was enough time to establish one of the major predictions of the theory. The amount of helium is a prediction of the first three minutes of the Universe, an idea first established by George Gamow. A Ukrainian refugee to the United States in the 1930s, Gamow made major contributions to nuclear physics and was also famous for his popular science books. A student of Alexander Friedmann, co-discoverer of the theory of the Big Bang, Gamow was an early pioneer of inserting that theory into mainstream cosmology. He realized, in the late 1940s, that here was the ideal furnace where helium may have been synthesized.

The Ashes of Creation

Solar power, the result of the release of thermonuclear fusion energy, is a key factor in establishing life on Earth. The principle behind thermonuclear reactions is that energy is produced when hydrogen fuses into helium, the second lightest element. The sum of the whole is less than the sum of the parts. Energy is released.

This same process occurs slowly but stably over billions of years in the cores of stars. Here the temperatures of tens of millions of degrees are high enough that thermonuclear fusion occurs. Thermonuclear fusion powers the Sun. Some of the helium we observe in the Sun was made in preceding generations of stars. Most was produced long before the Sun formed, in the first minutes of the Universe.

The history of helium is unlike that of nearly all the heavier elements. Most heavy elements were made in the centers of

FIGURE 5. The giant, elliptical-shaped galaxy Messier 87. In the center of the galaxy Messier 87, some 55 million light-years distant, is the first supermassive black hole to ever be imaged. This image was taken with one of the world's largest optical telescopes, ESO's Very Large Telescope, whose mirror diameter is eight meters. This giant galaxy contains exclusively old stars and is thought to be the host environment for the growth and nurture long ago of supermassive black holes.

Image credit: David Malin, © Australian Astronomical Observatory, http://messier.obspm.fr/Jpg/m87.jpg.

stars and released when the stars died, often in violent explosions. However, helium could not have been made in sufficient quantity in the stars. It had an earlier origin.

The Universe was once so hot and so dense that it resembled conditions in the center of the Sun. The thermonuclear fusion of hydrogen occurred in the first minutes of the Big Bang. The mass difference between hydrogen and helium is released as nuclear energy. The only way to see this energy directly would be in the form of the ghostly elementary particles called neutrinos that are emitted prolifically in nuclear reactions.[3]

We have detected solar neutrinos, which typically have energies of millions of electron volts. Detection is aided by their high energy. Solar neutrinos trigger reactions in vast deep underground vats of purified heavy water. The neutrinos scatter off of neutrons and produce energetic electrons and muons. Their propagation produces infinitesimal flashes of blue light. These are measured with large arrays of photon detectors.

A sea of neutrinos from the first minutes of the Big Bang is also predicted from everywhere in the sky. But so far such neutrinos have been too elusive to detect, as they are very cold. The cosmic neutrinos have energies of less than one-thousandth of an electron volt. Neutrino telescopes typically detect neutrinos from distant astrophysical sources with energies of hundreds of giga-electron volts. Fortunately, there is another fossil of thermonuclear reactions in the early universe: the helium produced in the first three minutes. The only solution to the mystery of the origin of helium is that it originated in the Big Bang itself, during the first minutes of creation. Only then was the matter hot enough for thermonuclear reactions to have occurred. During the first minutes the Universe went through a phase when it was as hot as the center of the Sun, and about as dense.

Remarkably, the predicted amount of helium produced in the Big Bang matches what we observe. Wherever we look, once we correct our measurements for the small contributions by the stars, we inevitably find the same helium fraction. It is a universal number. The detection of helium in the amount predicted during the first minutes is strong evidence for its origin in the remote past, long before there were any stars or galaxies.

There is another nuclear fossil, deuterium, or heavy hydrogen. We detect deuterium in the interstellar medium, but it is destroyed by stars. So stars are not a source of deuterium. The only possible source, again, is the Big Bang. Deuterium is made in the first minutes of the Universe and is a by-product synthesized on the way to producing helium. Enough is left over— about one deuterium nucleus for every 1,000 helium nuclei—to account for the observed deuterium. The amount that we observe directly matches the predicted fraction. So we have confidence in its origin in the first three minutes.

One remarkable consequence emerging from the helium prediction is that the Universe should still retain a cold fossil glow, despite the 14 billion years of expansion and cooling. A curious historical footnote is that, in 1952, long before the discovery of the cosmic microwave background, one of Gamow's graduate students, Ralph Alpher, predicted the temperature of the Universe today to be about 5 degrees Kelvin. But his prediction was soon forgotten.

A decade later, two US groups were in hot pursuit of fossil microwaves from the early Universe. These emissions would incarnate the faint glow from the beginning of the Universe. The group led by Robert Dicke knew what to look for and mounted an experiment in Princeton, New Jersey. But they were beaten in the race by the other group, two radio astronomers, Arno Penzias and Robert Wilson. In 1964, Penzias and Wilson discovered the relic glow by pure serendipity as they mapped the Milky Way in their work at nearby Bell Laboratories in Holmdel, New Jersey. The glow was totally isotropic radiation at a temperature of just 3 degrees Kelvin.

Astronomers have now made sky maps of the fossil radiation produced in the first weeks of the Big Bang, and the blackbody nature of the spectrum is confirmed. Astronomers have taken many measurements of the fossil deuterium and helium generated in the first minutes. It has been a triumph of pure thought. Theorists had deduced this hot, dense early phase simply by extrapolating the measured expansion of the Universe back in time.

The current evidence for the Big Bang is based on three pedestals—the three successive discoveries of the expansion of space, the abundance of helium, and the fossil radiation. This evidence makes an uncontestable science case. True, it is all circumstantial evidence, but no court would throw it out. It is hard to dispute the science.

Probing Perfection

Science advances by empirical test. We need to test the theory of the emergence of the fossil radiation from the beginning of the Universe. This radiation is a remarkably well preserved relic of a bygone era, and our only direct contact with the beginning of the Universe. But it needs further confirmation. Obviously the spectrum cannot be perfect. Energy is added as the early Universe expands, and matter is not an idealized smooth distribution. Fluctuations are growing, gravitational energy is being generated, and heating is inevitable to some extent. There should be infinitesimal deviations from a blackbody spectrum. Detection would help confirm the theory.

Perfection does not exist in Nature. There are always tiny flaws, perhaps visible only with a microscope. We learn from these flaws, but of course they may also seed our undoing. At one end of observable scales, genetic mutations powered human evolution; hence the immense race to understand genetics. Likewise, at the opposite end, we can use telescopes to look for cosmic flaws as we study the largest structures in the

Universe. It takes extreme sensitivity to look deep into the frequency distribution of the fossil microwave radiation.[4]

Theory tells us that the microwave background should be a perfect blackbody only in the absence of galaxies. Stars are made of baryons, but the photons in the fossil radiation outnumber the baryons by about 100 million. There are so many photons. Of course, today the photon energy is really low, but long ago a photon had more energy than the rest mass of an electron, and even, much earlier, the mass of a proton. So photons dominated the early mass budget.

Photons were so dominant in terms of energetics that any effects imprinted by the baryonic particles would be hard to detect. In principle, baryons could add scatter or radiate as long as there was some injection of energy, due to some early heating effects. These would add photons to the cosmic blackbody distribution. But if this occurred too early, no evidence survives.

In the first few months, the injected photons are absorbed and converted into blackbody photons. But add energy any later and the density has dropped enough that conversion is inefficient. Deviations remain. The energy distribution of the photons becomes distorted from that of a pure blackbody, albeit to a tiny extent.

The first such precision search for any deviations was made in 1990 by the COBE space telescope. It carried an instrument on board that compared the cosmic microwave background radiation to an internal ideal blackbody calibrator. No deviation was found over a wide range of frequencies. The scientists concluded that the cosmic background radiation was a perfect blackbody. They set a limit of one-hundredth of a percent on any possible deviations from a blackbody spectrum. This limit was only as good as the precision of their onboard calibrator.

But it was still remarkable. The most perfect blackbody ever produced had originated in the first months of the Universe.

In the following decades, scientists thought long and hard about improving on this result. The Universe is observed to be inhomogeneous. It contains stars and galaxies. The process of galaxy formation inevitably introduces deviations from the blackbody radiation early in the Universe. There must be such deviations if our theory of formation of structure is correct. Detecting these effects would be one of the ultimate tests of the theory.

However, the predicted effects are really small. We would have to improve on the COBE experimental limits by a very large factor. This is not easy to do on a free-flying satellite in space, and the required size is another challenge. An ideal environment for such an experiment would be in a cold dark crater on the Moon. The environment is tailor-made for a space experiment. There is no atmosphere that would block the desired signal at far-infrared frequencies. There is extreme cold in polar craters where the Sun is never high above the horizon. And above all, there is a stable surface on which to mount a large experiment.

The beauty of a lunar project is that we are going back to the Moon to develop infrastructure regardless of the cost. Once there, science projects are feasible on scales that are otherwise inconceivable. The Moon is the future for such high-precision telescopes.

CHAPTER 6

Our Violent Past

When tales of the cosmos are told, this period of ours may
always be recalled as that in which men first came to realise
what a violent universe we inhabit.

—NIGEL CALDER

The black hole is a completely collapsed object. It is mass
without matter. The Cheshire cat in Alice in Wonderland
faded away leaving behind only its grin. A star that falls into an
already existing black hole, or that collapses to make a new
black hole, fades away. Of the star, of its matter and of its
sunspots and solar prominences, all trace disappears. There
remains behind only gravitational attraction, the attraction of
disembodied mass.

—JOHN A. WHEELER

The Seeds of Violence

The first stars preceded the galaxies. Stars formed in ever-
growing accumulations until the first galaxies were assembled.
These are observed in a display of cosmic fireworks a billion
years after the Big Bang. Unfortunately, these are so far away
that only with our largest terrestrial and space telescopes can
we begin to barely glimpse these glories of the past.

But we know that when massive stars die, they form black holes. Our galaxy is teeming with these stellar remnants. They are observed as matter is tidally ripped from a binary companion star. The gaseous debris pours into the black hole. It heats up dramatically and swirls around at turbulent velocities approaching the speed of light. The end result is a luminous source of X-rays.[1] To detect cosmic X-rays, a space telescope is essential. Most of the X-rays from cosmic sources, fortunately for us, do not penetrate the Earth's atmosphere. Several different X-ray telescopes have measured profuse amounts of X-rays from stars and galaxies, and especially from the hot gas that pervades clusters of galaxies.

An important source of X-rays is the accretion of gas onto black holes. Typically, a compact object, often a black hole, has a close companion star whose atmosphere is overflowing. The gas falls onto the black hole, heats up in the intense gravity field, and prolifically emits X-rays. We monitor black holes in binary systems in our galaxy and in many nearby galaxies. Often the X-rays flare up, as the flow of gas can be erratic. We know that a black hole is a very compact object; otherwise, we would not be seeing the X-rays. The object is often too massive to be a compact star such as a neutron star. The most massive neutron star is about twice the mass of the Sun. Our only option is that the object must be a black hole.

This is very strong evidence, but it is indirect. We infer the presence of black holes that weigh a few times the mass of the Sun. There is just one way to detect black holes directly. The ultimate window is gravity waves. When two black holes collide and merge, space is shaken up. The vibrations in the gravity field generate a huge burst of gravity waves. Gravitational wave telescopes search for traces of the formation of black holes in remote galaxies. And many black holes have now been discovered. We will see that the ultimate future of gravitational wave astronomy may lie on the Moon.

Predicting Black Holes

In a remarkable demonstration of the power of scientific reasoning, black holes were predicted long before our modern technology allowed their detection. The story begins in the seventeenth century when Isaac Newton, a young Cambridge don, was taking refuge from the Black Death, an outbreak of bubonic plague that struck England for the last time in 1665, killing some 15 percent of the population. In his family's countryside garden, Newton became convinced that the force that explains why an apple falls from a tree also controls the orbit of the Moon. The gravity of the Earth steers the motion of a huge mass, the Moon, but equally it guides a small mass, the apple.

Gravity is a universal force, Newton argued. His theory turned out to be immensely successful in helping astronomers understand the orbits of the planets. The ocean tides were explained by the competing effects of lunar and solar gravity forces. The orbits of the nearby stars could also be explained.

But there were tiny flaws in Newton's theory. It eventually became clear that something was missing from it. One planetary orbit that had been well studied for centuries showed an anomaly: the advance of the perihelion of Mercury's orbit around the Sun. Newton's theory gave the wrong answer because it predicted a perihelion advance that was only two-thirds of the observed value. Something more was needed.

Albert Einstein rose to the challenge in 1916. Before he had even procured his first academic appointment, Einstein developed a new theory of gravity, which he called general relativity. Explaining Mercury's perihelion advance was a key motivation for Einstein, whose radical theory was verified within three years.

Two solar eclipse expeditions monitored the bending of light rays from background stars as they passed close to the Sun. The darkness of the total eclipse enabled the UK astronomer Sir Arthur Eddington, an advocate of Einstein's new theory, to measure the positions of stars at the limb of the Sun. The apparent star positions shifted slightly, as predicted by the bending of light rays by the gravity of the Sun. The experiments were a resounding confirmation of Einstein's work.

However, the new theory of gravity was only verified within the limit of relatively weak gravitational fields. More was to come, but only after a century had elapsed. Einstein predicted that when gravity is strong, a black hole must form. (He did not call it a black hole; this name came later.) The black hole is a region where gravity is so strong that light cannot escape. It is bounded by what physicists call a horizon. It is the point of no return for any material object or particle, and even for light. Einstein thought that such singular objects should not exist in nature.[2]

Nothing from within the horizon of a black hole escapes. Matter may shrink into a singularity in the center, but the outside observer will never know. In fact, light rays are strongly curved as the horizon is approached. This renders a cloak of invisibility at the horizon. The distant observer would see only a dark shadow. The black hole is silhouetted against any radiating matter that is swirling around and orbiting the black hole just outside the horizon.

The term "black hole" was coined by the general relativity expert John A. Wheeler in 1967. Half a century after Einstein's prediction, Wheeler became fascinated by collapsed stars, from which light cannot escape. He was motivated by a recent discovery of radio-emitting pulsars, the most compact stars ever detected. These objects have to be incredibly compact. Pulsars bleep, emitting regular radio blips every second with remarkable regularity. Indeed, a pulsar can exceed the accuracy of atomic clocks, the most accurate measures of time on short timescales and the most precise timing devices we have devised.

Pulsars were a total surprise to astronomers. They had to be a few kilometers in size, yet they had the mass of the Sun. Their extreme compactness made them different from any known stars. They had to be what we call neutron stars. Yet if neutron stars existed, then more massive compact stars had to collapse even further under the relentless tug of gravity as they aged and burned out.[3]

Wheeler was inspired by the discovery of compact stars. He inferred that black holes should actually exist. With sufficient gravity, nothing could stop their collapse. Einstein's pioneering conjecture was finally confronting reality half a century later. There had already been suspicions that black holes existed. The concept of black holes echoed earlier ideas about compact dark stars by two eighteenth-century natural philosophers, English geologist John Michell and French mathematician Pierre-Simon Laplace. The new innovation was the astrophysical take on black holes. The radius within which light is trapped is the event horizon of the black hole. No events within this region can be seen far away. For the remote observer, this is the apparent size of the black hole.

Dark Stellar Relics

Astronomy provides us with many remarkably compact stars. These are white dwarfs, the end points of Sun-like stars. The ultimate in stellar compactness is a neutron star, formed by the death of a more massive star. A white dwarf is a few thousand kilometers across, while a neutron star is only some 10 kilometers. All such compact stars are limited in mass to a couple of times the mass of the Sun. Anything more massive must collapse to a black hole.

As they begin to exhaust their supply of fuel, stars swell and shed most of their original mass. Stars that initially had masses less than about twenty-five solar masses end up as neutron stars. These are made of the most extreme form of matter. The nuclei are so highly compacted together that the entire neutron star is essentially at nuclear density. It is effectively one humongous neutron.

We know that a star that began its life with more than about twenty-five times the mass of the Sun cannot lose enough mass to end up as a neutron star. It is simply too obese and destined to collapse under its own weight. It contracts into a compact object weighing more than three solar masses. The end point must be a black hole—there is no other option. This is how most black holes formed.

Most black holes are produced from the deaths of massive stars that have spent most of their short lives burning hydrogen for thermonuclear fuel. They collapse to black holes once their supply of nuclear fuel is exhausted. Most massive stars turn out to be in binary systems. They have close companions. The orbital velocities and separations depend on the masses of the pair of orbiting stars. The orbital parameters provide us with a robust way of measuring black hole masses.

Let's consider a binary system, that is, two massive stars in orbit around each other. Most massive stars are born as binaries. Many are born over the history of the Milky Way. The more massive binary partner dies first and forms a black hole. As it ages, the companion star swells up. This is the beginning of its death throes as its core turns into a runaway fusion reactor.

Even the Sun will swell into a red giant star some four billion years from now. The companion, now a giant star, overflows onto the black hole companion, which prolifically emits X-rays. Our galaxy is peppered with luminous X-ray sources. Many are inferred to be accreting black holes.

The stellar partner in turn dies and forms a companion black hole. The orbital separation of the pair of black holes slowly shrinks as gravitational radiation is emitted. The circling black holes enter a macabre dance. Energy is lost as the orbit shrinks ever faster and the gravitational pull between the black holes becomes ever stronger. Once they are too close, there is no way back. The curvature of space is thoroughly shaken up as the black holes merge, combining to form a more massive black hole. The resulting vibrations in space and time generate a final burst of gravitational waves.

Einstein recognized that strong ripples in the gravitational field should escape to any distant observer. He predicted that gravitational waves are emitted as a consequence of gravitational collapse. Gravitational waves are the unique witness of the black hole merger. Detection would be the ultimate test of gravity, but a century passed before observations caught up with theory.

Black Holes Observed!

Einstein taught us that mass curves space. Space is curved by the presence of any mass, and more so by black holes. The curvature of space is shaken up by the transient passage of the waves produced as black holes form. It is the quivering of space that generates gravity waves. Energy is lost, and the black holes eventually merge together with a final huge blast of gravity waves. The trumpet sounds. It is here that the final adventure begins.

FIGURE 6. The M87 supermassive black hole in polarized light. The first image of a black hole was made with the Event Horizon Telescope in 2019. In this new view of the black hole in M87, the lines mark the reconstruction of polarized light, which is related to the magnetic field around the shadow of the black hole and is key to explaining how the M87 galaxy, located 55 million light-years away, can launch energetic jets from its core. The EHT collaboration linked eight telescopes around the world, including ALMA (the Atacama Large Millimeter/submillimeter Array) and APEX (the Atacama Pathfinder Experiment) in northern Chile, to create the EHT, a virtual Earth-sized telescope.

Image credit: The Event Horizon Telescope (EHT) collaboration, https://science .nasa.gov/m87s-central-black-hole-polarized-light.

But the signal at the Earth is weak. For a remote observer, the amplitude of the shaking of space is microscopic in scale.

The black hole merger events are so feeble because they are rare and so far away. In our own galaxy, we expect one such event in a million years. So we need to monitor millions of galaxies to catch a single black hole merger. By searching over vast volumes of space, we have a chance of finding a handful of transient events. And we need to be alert, attentive all of the time. The final gasp of gravity waves takes seconds or less, and they are easy to miss. Detector development took several decades. Finally, in 2016, physics was shaken up by the long-anticipated discovery of gravitational waves from the merging of a pair of black holes.

The signal that we received in our terrestrial detector was indeed very weak. Experimental confirmation turned out to be difficult. Gravity waves curve space infinitesimally, but in a time-dependent and directional way. Early attempts used metal bars, but these proved to be too insensitive. Any measuring rod suffers an infinitesimal change in length. The predicted change in length is equivalent to measuring the width of a human hair relative to the distance to the nearest star. In essence, monolithic bar detectors were too small. To go beyond bars a challenging experiment would be necessary.

Detection required the development of very special technology. The modern versions of gravity wave detectors measure the lengths of laser beams bounced off mirrors spaced several kilometers apart. Two perpendicular beams give two signals whose combination is used to reinforce the gravity wave signal by the principle of interference. A passing gravitational wave slightly stretches one arm while shortening the other. The observer looks for the increase in signal when the crests of two waves are in phase, and the cancellation in intensity when they are out of phase. One lines up the waves. Comparison gives the optimal sensitivity to the passing wave.[4]

The LIGO experiment was built to do just this. There are two sites, one in a former nuclear reactor reservation in Hanford, Washington, and the second in a humid pine forest in Livingston,

Louisiana. The 3,000-kilometer separation of the sites enables triangulation of the location of the gravity wave sources. The baseline has since been greatly extended to include a transatlantic component, the third detector, VIRGO, near Pisa, Italy. Comparing the beams tells us where in the sky the signal is originating. Hence we have a gravity wave telescope. But the positional uncertainty of any source is still large, amounting to tens of degrees.

We can measure the separations of the mirrors to accuracies of around the wavelength of the laser light. When a gravity wave passes by, it disturbs the lengths of the laser beams by the equivalent of less than one-trillionth of the width of a human hair. This extraordinary sensitivity is made possible by interferometry, which allows us to measure precision distances to an accuracy as small as the wavelength of the laser light.

The measured signal was actually predicted by the scientists. It is an application of Einstein's theory. Still, it was an immense relief to finally arrive at the long-promised land of gravitational wave astronomy. Theory shows that as the black holes approach, the gravity wave frequency rises with increasing intensity. Orbital velocities speed up as the orbit shrinks. We can view an eerie dance of death that culminates in a final chirp as the two black holes merge. All of this occurs in a fraction of a second—exactly as predicted by computer simulations of the expected signal.

Gravitational waves are the ultimate detection of the formation of black holes. Previous astrophysical evidence was circumstantial. Black holes are one of the most remarkable concepts in physics. It was to take nearly a century for them to be directly detected. Nobel laureate Kip Thorne, rewarded for his role in the long-anticipated experimental discovery of black holes, writes: "Of all the conceptions of the human mind from unicorns to gargoyles to the hydrogen bomb, perhaps the most fantastic is the black hole: a hole in space with a definite edge over which anything can fall and nothing can escape; a hole with a gravitational field so strong that even light is caught and held in its grip; a hole that curves space and warps time."

The final blip of the signal measures the mass of the black holes because its duration tells us the effective size of the black holes. The first detection was a merger of two black holes that each weighed about thirty times the mass of the Sun. The wave form of the signal is so precise that, because the Universe is expanding, we can infer the distance to the black holes. The further the black hole is from us, the more the wavelength of any emitted signal is stretched by the expansion of the Universe. From the spacing of the laser beam wave crests, the distance to the source can be inferred.

We know that black hole mergers occurred far away, in remote galaxies in the depths of the distant Universe, because two black holes take a long time to merge, perhaps a few billion years. These events are too rare to be found in our galaxy. Hundreds of merging black holes have since been discovered. The black holes range in mass from about twice that of the Sun to 100 times the mass of the Sun. The gravity wave telescopes are continuing to detect many more black hole mergers. We are entering a new era of black hole astrophysics.

Unfortunately, the optical counterparts to gravity wave sources are few and far between. Astrophysical black holes are notoriously bare, so their mergers are not expected to have any electromagnetic counterpart. A major advance came in 2017 when the merger of two neutron stars yielded a gravitational wave burst, along with the associated optical glow. Neutron stars with lower masses have much shorter signals than black hole mergers, so are easily distinguishable. A delayed gamma ray flash was detected. The electromagnetic signal yielded a wealth of information about neutron star mergers, the role of radioactive powering of the flash, and in the rare elements ejected into the interstellar medium. We even have a new name, kilonova, for the phenomenon. Since then, other neutron star mergers have been identified, and we continue to locate black hole merger events only within large patches of the sky.

Most excitingly for the future, we will be able to study the merging history of black holes by building a lunar gravity wave observatory. As the orbital timescale shortens, the gravity wave frequency rises. We will be able to follow many orbits of the infalling black holes as they come together for the final plunge. We will study their approach from first capture. Most excitingly, the distance of the Moon will allow triangulation of the signals to give accurate positions of the gravity wave sources. The Moon offers new horizons for gravity wave astronomy.

Where Massive Black Holes Are Lurking

There are two distinct varieties of black holes. Many formed from the death of stars; these stellar black holes are typically tens of solar masses. Others that formed from mergers of many smaller black holes and stars or from the monolithic collapse of clouds are monster black holes. This supermassive variety of black hole exists at the center of galaxies. They are observed to weigh from one million to 10 billion solar masses. Supermassive black holes formed long ago in the nuclei of galaxies, where conditions favored the growth of such monsters.[5]

Supermassive black holes can be highly active as they swallow gas and stars. The large energy release powers quasi-stellar radio sources known as quasars. Discovered in 1964, they were identified as extremely powerful, pointlike radio sources. When the optical counterparts were detected, quasar emission was found to feature both a continuous glow and spectral lines of distinct frequencies. The emission was produced by hot gas swirling around the black hole. The emission lines were found to be highly redshifted. The redshift was due to the expansion of the Universe. The observations showed that quasars are very distant and intrinsically very luminous.

We believe that quasars are supermassive black holes in the centers of massive galaxies. Debris rains in from stars that are torn apart by the immense gravitational tidal fields. Stars occasionally pass too close to the central black hole. The debris swirls around and heats,up owing to the immense tug of gravity. It settles into a compact disk orbiting the black hole. The disk accretes matter and radiates prolifically. Intense glows are generated in X-rays. Magnetic fields wind up around the spinning disk. Powerful jets of radio emission are magnetically accelerated and shot out along the axis of the disk, the path of least resistance.

Although quasars are the most luminous objects in the Universe, they are also highly compact. The luminosity of 1,000 Milky Way galaxies is generated within a region that is light-hours across. The Milky Way galaxy, for comparison, produces most of its starlight within a radius of some thousands of light-years. The only viable explanation of a quasar appeals to prolific energy release as debris from disrupted stars accretes around and feeds a supermassive black hole. In many cases, the mass of the black hole is inferred to be billions of solar masses.

Closer to home, a somewhat smaller massive black hole lurks in the center of our own galaxy. The Milky Way's major black hole is located in the direction of the constellation of Sagittarius. Studies of orbiting stars have allowed a precise mass determination. The stars can be directly resolved in our galactic center at infrared wavelengths. The orbital motions and speeds of stars around the black hole have been followed for more than a decade. The massive black hole at the center of our galaxy is found to weigh four million solar masses.

Our massive black hole happens to be a bright source of radio waves. It is one of the brightest in the sky, hence its name, Sagittarius A. Despite the strong radio emission, it is not very active today. The associated X-ray emission is also very low. Very little fuel is needed to maintain its low level of X-ray emission. We infer that Sagittarius A is accreting very little gas from its surroundings.

In the past, the situation was very different. We know that Sagittarius A once experienced outbursts of violent activity. About 10 million years ago, a giant explosion occurred that left traces in the gamma ray sky. The Fermi gamma ray satellite telescope image has revealed giant bubbles of gamma ray emission that extend around the galactic center out to hundreds of light-years. This discovery attests to the violence of the early outbursts from our central black hole.

We peer back in time with the world's largest telescopes to study the first quasars in the universe. The data we have collected suggest that supermassive black holes formed more or less coevally with the first galaxies. Both are present a billion years after the beginning. We observe the effects of black holes on their environment, effects for which we have no alternative explanation. Most striking are the vast outpourings of energy by quasi-stellar radio sources from a region only light-years across. These drive fast flows of gas out of the galaxy. The host galaxy loses its gas reservoir. It ages overnight and is destined to be a massive elliptical galaxy, full of old stars.

We can now do better thanks to gravitational wave astronomy. We can peer into the very center where the black hole is still growing. We can probe the depths of the monster.

The Emergence of Monster Black Holes

Essentially all galaxies contain central, massive black holes.[6] Some black holes weigh billions of times the mass of our Sun. We do not know in any detail how the monstrous black holes formed. Our best guess is that they grew by the mergers of many smaller stellar black holes. Here is where gravitational wave astronomy can enlighten us.

Another intriguing possibility is that black holes grow by swallowing large amounts of interstellar gas that accretes into the centers of galaxies, or by swallowing stars that venture too close. Prolific amounts of X-rays are emitted by the stellar debris. Again, this is a target for the new generation of X-ray telescopes.

We need extremely large telescopes to resolve such tiny regions, ideally without the intervention of the Earth's atmosphere restricting the clarity of our view. This is one reason why lunar telescopes will play a key role in future exploration of quasars.

Black hole captures of stars and gas accretion are aided by the mergers of galaxies. These violent interactions stir up the gravity field. One consequence is that the innermost gas and stars lose angular momentum. This phenomenon is especially vigorous near the black hole. Material is directed inward to the central black hole, facilitating the fueling of the monster.

Astronomers find indirect clues. One is a connection between central black hole mass and the total mass of stars in the inner galaxy. The inner regions of galaxies are spheroid-shaped bulges of densely packed stars. These spheroids contain the oldest stars in the galaxy. The first to form, they were born within the first billion years after the Big Bang. Massive black holes and stellar spheroids are inseparable and must have evolved together.

Disks of galaxies formed later. Disks surrounded the spheroid stars and contained many young stars formed by the recent acquisition of gas. As the gas flowed inward, its spin resulted in the formation of a giant disk. The Milky Way has a disk, and the Sun is in the disk. It also has a bulge, and at the very center of the bulge is a black hole the size of four million solar masses. These results have two implications. First, we expect to find massive black holes a few hundred million years after the Big Bang. And indeed, black holes as massive as 10 billion solar masses are found when the Universe was a billion years old. Second, seeds—smaller black holes—are needed to accrete gas or merge to make the most massive black holes. Otherwise, the monster black holes could never be formed early enough.

If they started from typical stellar black holes, around ten or twenty solar masses, there would barely be time enough to grow the monsters we find. One intriguing discovery is that the most monstrous black holes were already present in the distant past. The most massive of these weighs in at 20 billion Suns and was already in action as a source of luminosity a few hundred million years after the Big Bang. How the seeds of monster black holes were formed is somewhat of a mystery.[7]

Most likely, supermassive black holes emerged after the Big Bang from the first generation of clouds formed, some of which contained a million solar masses of gas. These clouds are thought to be the sites of the first stars. Traces of hydrogen molecules would have provided the needed cooling. But many of these clouds might not have cooled sufficiently. Molecules are easily destroyed where black holes are lurking. Some clouds would have been too warm to fragment, but would still have lost enough energy to collapse. Some of these first clouds must have undergone monolithic collapse, bypassing fragmentation for the most part.

In this way, the massive black holes formed. We expect them to weigh in at tens of thousands of solar masses. These black holes are intermediate in mass between those formed directly by dying stars and the monsters detected in the centers of massive galaxies. They are the likely seeds of the monsters. And we have discovered them.

Here is how we test this hypothesis. We look for the seeds, and our best bet is dwarf galaxies. Many dwarf galaxies contain central massive black holes. These are intermediate mass black holes. Massive galaxies formed from the merging and accretion of many smaller dwarf galaxies. As dwarf galaxies merged together, their black holes sank into the centers of the resulting massive galaxy. The seeds were gathering. Then they merged to eventually form a supermassive black hole. Every massive galaxy most likely had such a lurking giant in its center. And most dwarfs might have had an intermediate-mass black hole. These are exceedingly difficult to observe, but many have recently been detected.

Seeing the Monster

The black hole continues to grow in mass. It accretes nearby gas clouds, which are everywhere in young galaxies. The mass doubling time for a typical massive black hole is about 50 million years. Gas clouds are captured and the central supermassive black hole, a quasar, is fed. Quasar eruptions continue sporadically. We have found that quasars were dramatically more abundant in the early Universe, when there was a plentiful fuel supply of gas.

Black holes become inactive when they are not being fed. They are reactivated only when fresh gaseous fuel is provided. Supermassive black holes at the centers of nearby galaxies are dormant giants. They were active in their youth as quasars, long ago. At that time, galaxies formed stars prolifically, but today most massive galaxies are red and dead. There are only old stars that are cool and red. Massive galaxies are mostly dead because, in the absence of a supply of molecular gas clouds, young stars are not forming.[8]

There is an intricate link between galaxies and central black hole growth. Massive galaxies all seem to contain supermassive black holes. Black holes cannot overfeed and become obese. The violent effects associated with black hole feeding trigger outflows of gas. If a black hole were to get too massive, the outflows would be enhanced and any residual gas would then be mostly blasted away with even greater efficiency. Outflows succeed in clearing out the interstellar medium in the host galaxy. Star formation in the forming galaxy is quenched. A universal ratio of black hole to stellar mass is maintained. The more massive a galaxy is, the more massive is its central black hole.

This is self-regulation at work. Remarkably, we observe such a relation between the mass of the galaxy and the mass of the central black hole from the smallest to the most massive galaxies. This tells us that old stars and central black holes control each other. Both formed early in the history of galaxies. The radiation from the central black hole limited the amount of infalling gas that could fragment into stars, and the inhibiting accretion of gas in turn limited the black hole mass. Self-regulation leads to a successful dietary regime.[9]

The first direct clues to supermassive black holes came from radio astronomy. The ultimate proof of the existence of supermassive black holes was in the form of radio wave imaging of the black hole shadow. The trapping of light from the accretion disk around the black hole leaves an imprint. Silhouetted against this is the shadow of the giant black hole. Light from the orbiting of matter around the black hole cannot escape and thus leaves a nearly circular shadow.

The first chosen target was at the heart of the giant elliptical galaxy Messier 87. The closest really massive galaxy to us, Messier 87 is some 50 million light-years away in the constellation of Virgo. The size of the shadow measures the horizon of the black hole. After years of painstaking preparation, M87 finally was observed in 2019 by a worldwide assembly of radio telescopes acting in unison. The Event Horizon Telescope utilized simultaneous data taken by eight different radio telescopes in diverse locations ranging from the South Pole to the United States to Spain.

The aim was to combine the different signals from each telescope to produce a single combined image. The technique is known as interferometry because interference patterns are generated by the telescope signals. The observation makes use of signals taken from telescopes that are thousands of miles apart. Interference patterns are produced when the radio waves from the different telescopes are combined in phase—that is, peaks with peaks, troughs with troughs. We achieve the combined signal thanks to the exquisite timing of ultraprecise clocks. This technique allows consolidation of the individual telescope signals into a single image.

Coordinating the phases of the radio waves enables the global network to act as a single giant telescope that is thousands of kilometers across. The angular resolution achieved was 25 millionths of an arc-second, determined by the largest separation of the radio telescopes. That resolution corresponds to the horizon scale of the black hole, which is about 0.001 parsecs. The black hole shadow is about two and a half times this size. The mass of the monster black hole in Messier 87 was found to be 6.5 billion solar masses. And remarkably, a lopsidedness in the surrounding glow was determined to be due to the pull of the black hole's spin on the surrounding space-time. This effect has been used to demonstrate that the black hole is rotating at a significant fraction of the speed of light.

The Search Goes On

The search for supermassive black holes continues. The ultimate window on forming and observing massive black holes is gravitational waves, which are emitted as supermassive black holes orbit each other. Eventually the black holes merge in a final gasp of gravitational waves.

Black hole mergers are predicted to occur as galaxies merge. The frequency of the gravity waves produced is much lower than that for stellar black holes because the orbital timescales involved are much longer. Stellar black hole mergers occur over milliseconds and generate gravity waves with kilohertz frequencies. Supermassive black hole mergers take much longer.

Supermassive black holes merge on longer timescales that increase as the black hole mass increases, because the horizon size is proportional to the mass of the black hole. It takes as long as ten minutes for the massive black hole merger to occur. This time span corresponds to producing gravity waves with frequencies of only a millihertz. The massive black holes orbit each other many times before merging, traversing millions of miles at near-light speed. Detecting such low frequencies can only be done from space. We will need to bounce a laser beam many times between satellite detectors, each, say, one million miles apart. Only then can the needed sensitivity be achieved to detect the long wavelengths of hundreds of millions of miles that correspond to the low frequencies of the gravity waves produced as the supermassive black holes finally merge.[10]

This is exactly the plan for LISA, a space interferometer to be launched by ESA in 2037. LISA consists of three satellites flying in formation far from the Earth. The satellites will fly in an equilateral triangular formation, with 2.5 million kilometers on each side. They will orbit the Sun about 50 million kilometers from the Earth. Each satellite will contain a test mass whose position will be carefully measured. The precisely calibrated sides of the three-satellite triangle will provide the perfect rulers for the gravity wave probe.

The test mass must experience zero drag for the experiment to work. Its path through space has to be controlled only by gravity. It cannot be affected by nongravitational forces on the satellite, such as drag from the impact of the solar wind or radiation pressure from the Sun. This is accomplished by placing the test mass in a vacuum cavity inside the spacecraft. The cavity wall will suffer these forces, and the outer shell will be repositioned by thrusters every time the spacecraft wobbles. In this way the enclosed test mass in the cavity will feel only the local gravitational field. Its isolation is designed to probe the effects of passing gravity waves.

The distance between the satellites will be monitored by lasers. The precision needed must be high enough to allow detection of gravitational waves. To achieve this goal, LISA will have to measure relative displacements with a resolution of one-billionth of a centimeter, or less than the diameter of an atom, over a distance of one million kilometers. By launch time, our technology promises to be up to the task.

A Lunar Gravity Wave Observatory

There is another way to detect gravity waves. An isolated bar rings like a bell when impacted by a gravity wave. The ringing is weak, and the effect is small. Terrestrial attempts to build bar detectors have all failed so far. However the Moon provides another route. It promises to be the ultimate gravity wave detector. Seismologically, it is far quieter than the Earth. So why not build a gravity wave telescope on the Moon? The Moon is like a giant bar detector. Here is how this might work.

One idea is to install several seismometers in different regions of the lunar surface. A seismometer is a pendulum-like device that measures tiny vibrations on the ground. It is used to monitor earthquake activity. Apollo astronauts placed seismometers on the Moon, where they were monitored for several years. Lunar quakes, which are mostly minor, are not attributable to tectonic activity. In fact, the Moon has a solid core. There is no equivalent of the floating tectonic plates whose motions and collisions cause terrestrial seismic activity.

We saw earlier that slow thermal contraction of the Moon's crust and tidal disruptions of the mantle by the pull of the Earth and the Sun lead to stress accumulation, which is relieved by occasional quakes. Meteorite impacts also play a role in perturbing the crust. Yet overall, lunar seismic activity still seems to be millions of times weaker than on the Earth. This makes the Moon an ideal site for gravitational wave astronomy because it allows us to look for slight vibrations induced by passing gravitational waves. The idea would be to coordinate signals from seismometers hundreds of kilometers apart, looking for any common disturbance.

The science prospects are unique. We would be looking at frequencies intermediate between those reached by LISA in space and LIGO on the ground, and we could begin to explore a new range of massive black holes—those that are intermediate in mass between the stellar relics and the supermassive monsters in the nuclei of massive galaxies. Understanding these intermediate mass black holes, which are the elusive seeds conjectured to bridge the gap between the large and small black holes, seems to be an essential element for understanding the existence of supermassive black holes in the very early Universe. A lunar gravity wave telescope will provide the clues by observing the inevitable mergers of such black holes as they generate flashes of gravitational waves.

We will fill the gap between 10-kilometer and million-kilometer baselines. The gap determines the frequencies at which we can detect black holes as they inspiral and finally merge together. As they orbit ever faster, they emit gravitational waves at higher and higher frequencies, culminating in a final chirp. Bridging the frequency gap without interruption will enable us to study this dance of death in gory detail, from beginning to end. The lunar surface allows for the operation of gravity wave detectors that will plug this gap. Seismometers monitoring the vibrations of the Moon are ideally suited for this purpose. We will finally be able to study the birth pangs and death throes of massive black holes.

There would be another bonus. In combination with terrestrial gravity wave detectors, triangulation could locate the direction of the gravity wave source. We should be able to identify the locations of gravity wave sources with a hundred times the angular precision available with terrestrial detectors. Gravitational wave astronomy would come of age.

CHAPTER 7

Are We Alone?

In space there are countless constellations, suns and planets;
we see only the suns because they give light; the planets
remain invisible, for they are small and dark. There are also
numberless earths circling around their suns, no worse and no
less than this globe of ours. For no reasonable mind can
assume that heavenly bodies that may be far more magnificent
than ours would not bear upon them creatures similar or even
superior to those upon our human earth.

—GIORDANO BRUNO

I'm sure the universe is full of intelligent life. It's just been too
intelligent to come here.

—ARTHUR C. CLARKE

Humanity's Greatest Question

Where are they? This question about the existence of other intelligent beings in the Universe poses immense challenges. Do
we infer that we are alone in the Universe? We speculate on
answers, but only direct searches can provide them.

During a summer visit to Los Alamos National Laboratory,
around 1950, nuclear physicist Enrico Fermi is said to have
asked this famous question. The setting was a luncheon with his

colleagues. One scientist later recalled that they were on their way to the luncheon and bantering about a *New Yorker* cartoon depicting aliens stealing public trash cans from the city streets. The subject was later dropped, but in the midst of lunch Fermi suddenly shot off his question. He was expressing his surprise over the absence of any signs of the existence of other intelligent civilizations in the Milky Way galaxy.

Many potential resolutions to the so-called Fermi paradox have been suggested over the years. I'll describe some of them in this chapter. Researchers still have not reached consensus on which one, if any, is likely to be correct, but any reasonable set of assumptions tells us that a technological civilization could have reached every corner of the entire Milky Way, and within a time much shorter than the age of the solar system. We have not yet encountered any alien visitors from outer space. Nonetheless, the question of whether we are alone in the Milky Way, or in the Universe at large, remains one of the most intriguing questions facing modern humans.[1]

Our solar system formed halfway through cosmic history, some 4.6 billion years ago. That's fairly late in the history of the Milky Way. Much could have preceded our arrival. Numerous Sun-like stars are billions of years older than our Sun, and many host planetary systems that are similarly very old. The odds of finding evidence for life in other solar systems are completely unknown.

Nevertheless, the search continues. Astronomers are observing stars throughout our galaxy for evidence of Earth-like planets, seeking potential signs of life on remote planets that resemble the Earth in atmosphere, composition, and climate. With the next generation of large telescopes in space, we will soon be able to study the nearest exoplanets in detail. These telescopes will provide clues to possibly intelligent life such as elevated levels of oxygen. A substantial fraction of atmospheric oxygen is a necessary ingredient for life as we know it. We will seek additional biological signatures of increasing complexity.

Riding on answers to the question "Are we alone?" is nothing less than our claim for being special in the cosmos. We shall never know how special we are unless we search for other living beings, and the ultimate platform for that search, I argue, is the Moon.

What Is Out There?

Attempts have been made to estimate the probability of the existence of intelligent life elsewhere in the Universe. They are complicated, however, by our ignorance. There are too many unknowns when it comes to assessing the odds for intelligent life.

Beyond the question of whether extrasolar life even exists, there are enormous uncertainties about its emergence, evolution, and survivability. In spite of remarkable progress toward producing life in the laboratory in recent years, beginning with a tepid soup of organic matter, the precise origin of life—the dramatic transformation from chemistry to biology—remains a mystery. Similarly, even though Darwinian evolution has proven to be an enormously successful paradigm for understanding the diversity of life, the central fact is that life on Earth is the only example of life we have so far. Based on such limited statistics, the probability of life elsewhere remains a source of extreme unpredictability.

One experimental goal is to search for life elsewhere in the solar system. Two promising environments are Mars and Jupiter's fourth-largest moon, Europa. Mars is arid and apparently lifeless, but traces of water have been found there. From the

FIGURE 7. Orbiting exoplanets are visible in the star system nearest to our solar system. The two bright stars are Alpha Centauri (left) and Beta Centauri (right), both binaries. The faint star in the center of the circle, southeast of Alpha, is Proxima Centauri. This intensely red, smaller, and less bright star is a distant third element in a triple star system whose close main pair form Alpha Centauri. Because of their Earth-like characteristics, they are prime targets in searches for signs of extraterrestrial life.

Image credit: Skatebiker at English Wikipedia, CC BY-SA 3.0, https://commons .wikimedia.org/w/index.php?curid=46833562.

dried-out lake beds and river canyons we infer that there was abundant water in the past as well as an oxygen-rich atmosphere. Excavation may reveal traces of fossil life, or at least evidence of microbes. This is a central goal of future exploration.

Likewise, the oceans below the kilometer-thick icy crust of Europa are a potential environment for aquatic life. The tidal gravitational forces of Jupiter stir and heat the deep interior of its moon, providing internal energy that maintains the liquid oceans. ESA's JUICE mission will give us detailed imaging after its expected arrival at Europa in 2029.

Another promising candidate with subsurface salty oceans is one of Saturn's moons, Enceladus. Here icy plumes of ice and water are ejected through cracks in the icy surface. The Cassini probe analyzed the composition of the liquid water and found that the oceans must be rich in organic molecules. These are a primary ingredient for synthesizing amino acids and proteins, the fundamental components of the simplest forms of microbial life.

Discovering traces of life elsewhere than on the Earth would provide support for the idea that life on our planet may not have been just a one-off accident. Establishing evidence for the ubiquity of life would help us better understand the important role of serendipity throughout the history of life on Earth. For example, Earth is blessed with a relatively large moon that has helped to stabilize the climate. The asteroid belt may have been responsible for mass extinctions. It is also conceivable that the huge conflagration triggered by an asteroid collision may have helped to seed life. Organic materials may have been part of the asteroid debris.

Our Sun may be in a preferential location for habitability. The location of the solar system within a minor spur off one of the two main arms of the galaxy, relatively far from the galactic center, has shielded us from the potentially sterilizing effects of gamma ray bursts. These originate in highly magnetized, spinning neutron stars. Fortunately, we are probably far from where any potentially life-destroying events could take place.

We have no idea how complex organisms might evolve. There are many random evolutionary directions that lead nowhere. Darwinian evolution, which seems to have worked on the Earth, is often invoked, but for all we know it could have been an immensely improbable fluke. With one example, that of life on Earth, our statistics are nonexistent. Other planetary environments may be unstable, if not hostile. Intelligent life may be self-destructive. Perhaps life is even destined to recycle many times. There would never have been enough stability for life to achieve the maturity of, say, the million-year durations needed for advanced civilization to emerge. There would not be time enough to develop the technology that would allow interstellar travel.

On the other hand, intelligent life could have been inevitable, given the billions of years available, especially around other Sun-like stars. All we can do is search, since we cannot estimate the probabilities involved. Our search is hampered, however, by one problem: we don't really know what advanced life signatures we should be seeking. Intelligent life that could accomplish interstellar travel would inevitably be thousands or even millions of years ahead of us in technology and would clearly have had so much more time to evolve. Would we even recognize such creatures if they were in front of us? Invisibility cloaks would be a trivial technology for such denizens of the Universe, if indeed they exist.

Life in the Solar System

The topic of extraterrestrial life and how to detect it has become particularly timely. The mundane but essential approach is to look locally in the solar system for signs of life beyond the Earth. Three main ingredients are needed for life. First, liquid water oceans must be present to provide a solvent that allows molecules to synthesize. Next, there must be a nontoxic atmosphere to shield against life-destroying extreme ultraviolet radiation. Finally, demonstrating that not all radiation is bad, energy is required to aid molecules in synthesizing.

In the solar system, the Moon and the planet Mercury have no atmosphere. These are not promising environments in which to search for life. The atmosphere of Venus is toxic: raining sulfuric acid is not compatible with life. Phosphine was recently claimed to have been discovered in the Venusian atmosphere by its absorption against the radio emissions from the cloud layers. Phospine's only known source on the Earth is microbial activity, but it is also seen in the clouds of Jupiter and Saturn, where phosphine is probably not formed by microbes.

Venus is more similar to Earth, apart from its runaway greenhouse gases. Life would seem difficult to sustain in Venusian clouds. Volcanic activity on Venus offers another possible source of phosphine. Nor is that the only solution. A reinterpretation of the original radio features argues that the observed signal may equally be due to sulfur dioxide, a common gas in the Venusian atmosphere, The jury is out on what the possible detection of phosphine in the clouds of Venus actually means.

Mars is the best bet for seeking traces of fossil life. Mars once had a dense atmosphere, and its surface is shaped by water and wind-driven erosion that occurred billions of years ago. Elsewhere, the oceans of Enceladus, an icy satellite of Saturn, are conceivable depositories of microbial life. Other Saturnian satellites, such as Titan, the largest satellite in the solar system, are fascinating worlds to explore. Titan's methane atmosphere, however, would not be conducive to life as we know it. Methane can be produced organically, and there would be traces of carbon dioxide too. But astronomers doubt that this is the case for Titan, whose environment seems too hostile.

Yet Titan's pervasive methane haze serves as a prototype for astronomers searching for evidence of organic forms of life on young Earth-like planets. Conditions on the early Earth might have resembled Titan. About three billion years ago, the Earth had a methane-rich atmosphere. Then an explosion of life generated prolific amounts of carbon dioxide. This gas was broken down by ultraviolet radiation from the young Sun to produce our planet's present oxygen-rich atmosphere.

Exoplanets

Even if life tracers, such as microbial fossils, were found by our future missions to Mars or Europa, this does not really inform us definitively about the ubiquity of life. There is too great a risk of terrestrial pollution. After all, a few terrestrial meteorites have been identified as having a Martian origin.

We need to search farther afield than the solar system. Observations made primarily with the Kepler space telescope have shown that for every cool dwarf star in the Milky Way, there is a significant chance of finding at least one Earth-sized planet orbiting its so-called habitable zone. The habitable zone is the "Goldilocks" region that allows liquid water to exist on the planet's rocky surface because the planet is neither too far from the star, where it's too cold, nor too close, where it's too hot.[2]

Adding possible options as we extrapolate the Kepler statistics, we arrive at an expectation of more than a billion such Earth-like planets in our Milky Way galaxy. And adding M-dwarfs as likely exoplanet hosts sends this number through the roof. The Kepler telescope search statistics alone uncover thousands of promising exoplanets that are the closest and the easiest to study. Perhaps something interesting is lurking in this horde—unless, that is, the probability is exceedingly low that a habitable-zone planet would develop a technologically capable species of life. In this case we might need to search a million candidates or more. Even then, we will find no advanced biotracers if life is truly rare.

How rare would life have to be if we are truly alone? We can estimate that life would need to break out less than once on a billion trillion exoplanets. This is just the expected number of terrestrial-like planets in habitable zones in the observable universe. These odds are very low indeed, but perfectly compatible with our ignorance about the origin of life.

But perhaps this line of thinking is too pessimistic. Unless we are very unlucky, it is not unreasonable to conclude that Earth's humanity should not be the only technological civilization to have ever existed in the Universe. Certainly there has been lots of time. But we must also acknowledge that such advanced entities may be not only really rare, but simply very far away, perhaps beyond our current reach.

Having a chance to answer at least some of these questions is the prime reason why constructing bigger and better telescopes could bring immense rewards. We simply need to look farther afield. Key to our future efforts to extend our reach will be the Moon.

Looking nearby is certainly a useful pilot strategy. To more reliably assess the odds of life existing elsewhere, we desperately need more examples of Earth-like exoplanets. Let's start with the closest candidate. At a distance of 4.2 light-years from Earth, Proxima Centauri is the Sun's closest stellar neighbor. It is a red dwarf star, with a mass of only 12 percent of the Sun's. Its luminosity is 17 percent of the solar luminosity. The habitable zone is consequently much closer to the star than is the case for the Sun. Within this zone, an Earth-like planet orbits every eleven days. It receives an energy flux that is about 70 percent of what the Earth receives from the Sun. This is certainly enough to generate a viable biosphere.

We have a good place to start the search. Of course, the skeptic would argue that putting all one's eggs in one basket is a strategy doomed to failure. We will need to go much further. But we have to start somewhere.

Signatures of Life

To search more broadly for signs of life, we need to study the atmospheres of exoplanets, as many as we can access. We need to look for biological signatures. One good example of a signature of life is the Earth's own atmosphere, which glows with evidence of plant life. We know this because its light is reflected as "earthshine" when there is a crescent Moon. Earthshine contains absorption lines that, seen against the Moon's light, show evidence of the "red edge." This feature in the spectrum of light from the Earth's atmosphere stems from ultraviolet light being absorbed by green vegetation and re-emitted at longer wavelengths. So we see a red absorption feature in the spectrum of the light. There are even seasonal variations in this red spectral edge, owing to the presence of vegetation.

The initial idea behind searches for biological signatures is to look for similar spectral features in the atmospheres of distant planets. We would measure them in absorption against the glare of their parent stars. We could even look for seasonal variations. So far we have not seen such features, but we have had very few accessible candidates.

Detection of the red edge will be a key goal of future telescopes in space. The Earth's atmosphere has restricted transparency for many of the most desirable spectral bands needed for biomarker searches. Only space telescopes have the necessary resolution and freedom from atmospheric limitations, though, as I argue, lunar telescopes are the ultimate future.

Searches for life focus on Sun-like and smaller stars because the vast majority of stars are smaller than the Sun. Typically known as M-dwarf stars, they weigh in at around one-third of the mass of the Sun. These red dwarf stars make up something like 70 percent of all stars in the Milky Way. Just like our nearest stellar neighbor, Proxima Centauri, a large fraction of these M-dwarfs may harbor planets.

Two factors make more massive stars less hospitable as life-supporting energy sources: they have shorter lifetimes and emit intense ultraviolet radiation. The biochemical processes necessary for life may require billions of years to unfold. Stars more massive than about three times the mass of the Sun, for instance, are likely to burn out before life has time to emerge and evolve. Yellow dwarfs like the Sun, as well as the more common red dwarfs, are more common and live much longer. As an added bonus, the planets orbiting and transiting these redder and dimmer host stars are easier to detect against their glow.

It is simpler to begin by generalizing our search to simply seeking evidence for life elsewhere. This strategy could encompass looking for something as simple as atmospheric traces of microbes—or at least, taking a step back, the conditions in which microbes might flourish. If we were to find that such organisms exist elsewhere, that would greatly bolster our search for more complex forms of life. There are three essential conditions for life to appear: We need a solvent such as water. We need organic material. And we need energy.[3]

Some form of liquid solvent is undoubtedly necessary if chemicals are to be transported into and out of cells. This is essential if molecules are to come into contact with one another to form long-chained organic matter. A liquid environment would also protect that organic matter from ultraviolet radiation. It is not entirely clear whether water alone can play that role. There may be alternatives, so we need to keep our options open.

To be detectable from a distance, life has to evolve to the point where it so dominates the planetary surface chemistry that it has significantly altered the atmosphere. Only then will a planet with life give itself away through chemical biosignatures. With sufficiently powerful telescopes, such biosignatures could be detected remotely.

Earth itself would probably not have been detectable as a life-bearing planet during the first billion or more years of its existence. Oxygen is an important biosignature, and it became a major atmospheric constituent at a relatively late stage, almost entirely owing to the simplest photosynthetic organisms that developed some two billion years ago. Microbial bacteria flourished, and the oxygen content of the atmosphere built up slowly.

The accumulation of large amounts of atmospheric oxygen is first traced by its effects on oxidizing iron-containing rocks. We find rusty deposits at the floors of the oceans. By age-dating rocks, we can infer when the oxygen-rich atmosphere first developed. After prolific oxygen production by biological sources, which made use of energy from sunlight, oxygen began to enrich the early terrestrial atmosphere. Only then did primitive multicellular precursors of life develop.

What *single* detectable biological signature might be considered the most reliable for indicating the existence of life on a sufficiently old, rocky planet in the habitable zone around its star? No single biosignature would be absolutely compelling, as there are multiple pathways to the most primitive life. An atmosphere that is *very* rich in oxygen—say, at a level of a few tens of percent—would probably be the most promising target initially. But we would need to follow up with complementary searches.

Nonbiological processes can produce oxygen in a planetary atmosphere. These include the splitting of carbon dioxide by intense solar ultraviolet radiation. As water vapor evaporates, hydrogen is lost. Yet only under rare circumstances could these processes create the high levels of stable oxygen enrichment needed for life. One could imagine a case in which there is no need for a biological source: volcanic activity producing prolific amounts of carbon dioxide that is exposed to ultraviolet radiation from the young Sun and dissociated into oxygen.

To argue for a biological origin of oxygen, we would need to see the presence of oxygen combined with other potential biological signatures. One possibility is the chemical consequences of an extreme thermochemical disequilibrium. Such a phase would be indicated through the simultaneous presence of oxygen and methane, which are not normally produced together. Methane is a key waste product of biological activity. What would also help would be a determination of the ratio of ozone to oxygen. Only with a combination of several chemical markers would the credibility of a life-based origin for the oxygen be significantly strengthened.

We will look for habitable surface conditions with atmospheric biosignatures, for instance, water vapor and greenhouse gases such as chlorofluorocarbons, ozone, and nitrous oxide. In the Earth's atmosphere, methane is produced by bacteria when cows belch and by plant decay in swamps. Other targets include organic hazes and the vegetation-caused red edge, an increase in reflectivity due to foliage structure. We will look for seasonal changes in atmospheric biomarkers associated with photosynthesis, including carbon dioxide variations. We will look for seasonal glints off the oceans. But we will need very large telescopes in spacelike conditions. That is only feasible on the Moon.

Radio Signatures of Habitability

Suppose an exoplanet survives these many obstacles to the generation of life.[4] All could still be lost in the typical exoplanet environment. Young stars are capable of dramatic outflows, and they also generate intense fluxes of cosmic rays and X-rays. The prospects of habitability are diminished on an exoplanet being blasted with intense stellar winds. Exoplanets could be sterilized as a result, dramatically reducing life options.

The Earth is protected against the solar wind and solar cosmic ray eruptions by its magnetosphere, which deflects incoming ionizing particles. If a magnetosphere is in place around a young exoplanet, it is protected against such environmentally destructive phenomena.

The physics of a magnetosphere is well understood, and we now have a way of detecting the presence of a protective magnetosphere around an exoplanet. Interaction with the solar wind generates low-frequency radio signals, which are often flares. On the giant planet Jupiter, a strong source of radio flares, the flares are often triggered by the passage of Jupiter's satellite Io, which has a magnetosphere. As the planet passes near its host star, huge currents are generated in the stellar magnetosphere. These trigger aurorae and radio emission. Similar low-frequency radio experiments planned for the lunar far side will monitor nearby exoplanets and be able to deduce the presence of magnetospheres.

Perhaps organic material can be resupplied to sterilized exoplanets by cometary infall. Comets are hypothesized to be an important source of terrestrial water as well as possible suppliers of organic matter to the young Earth. Whether this adds up quantitatively is less clear.

Overcoming the Glare

An excellent first step in the quest for signatures of extrasolar life in the relatively near future would be to search for planets with abundant atmospheric oxygen. This goal can be achieved by building large, ground-based arrays of relatively low-cost flux collector telescopes. Another approach would be to use the European Extremely Large Telescope, now under construction. This will have a light-collecting area equivalent to the size of several football fields. The world's largest telescope will be equipped with very high-dispersion spectrographs, which are essential for finding biological tracers.

First, however, there is a major hurdle to overcome. Dazzle from the light of the host star renders it impossible to see exoplanets. To directly visualize an exoplanet, we need to suppress the starlight down to one part in a billion. This is not easily feasible from the ground because our atmosphere scatters starlight. But even overcoming this obstacle alone will not be enough to allow us to seek definitive signals. We need to also look in spectral regions that are blocked by the Earth's atmosphere. Ideally, to do better, we need to go into space.[5]

We must go beyond the Earth because our atmosphere cuts off so much of the electromagnetic spectrum where hints of life are found. Infrared and far-infrared wavelengths must be exploited over the entire electromagnetic frequency band. A first step toward a solution is under development. The Nancy Roman Space Telescope is a wide-field infrared survey telescope to be launched in the mid-2020s. It will be equipped with a star shade, a device to shield the exoplanet from most of the light emitted by its host star.

Imagine doing astronomy near a streetlight. The night sky would not be very dark. But it is possible in space to get the necessary protection against the light from the host stars of the distant planets. In space the star is a bright point of light, and we can use a star shade to block out the starlight. A star shade is like a giant umbrella made of mylar and about the size of a tennis court. The shade is launched separately and then unfurled in space some tens of kilometers in front of the telescope. Both are in solar orbit a million miles from the Earth. Designed to block the starlight before it reaches the telescope, the star shade acts like an opaque occulting shield that covers the star but not the orbiting exoplanet.

Fortunately, the star's glare is lower in infrared wavelengths, and most of the stellar targets shine less brightly in the infrared. But we still need to block 99.999999999 percent or more of the light from the star. Doing so requires exquisite positioning of the star shade. Only then can we hope to resolve an orbiting rocky planet.

Our aim is to characterize distant Earth-mass planets and detect signals of biological activity. Because we expect these signals to be very weak, our best chance of isolating them is at longer wavelengths to detect a planet that is dimly shining by reflected sunlight. Space telescope sensitivity and contrast is optimized in the infrared. That's where we should look.

Even if we can control the stellar glare sufficiently, we can do this only for the nearest stars because we are limited by the telescope's resolving power. The space telescope's aperture and light-gathering power limit the number of detectable planets. We'll see how the Moon comes to the rescue.

Unless rudimentary life forms are extremely ubiquitous and detectable remotely, the probability of detection is not very high. Our best estimates are inherently highly uncertain. Most habitable-zone planets are unlikely to have an oxygen-rich atmosphere. Like Mars, they are too low in mass to retain their atmosphere.

Other hurdles must be surmounted. Perhaps the host stars have frequent life-destroying flares. Perhaps there is a paucity of the volatile elements required for synthesizing organic matter. The chemistry must be right for the emergence of life. There must be liquid oceans of water and an oxygen-rich atmosphere, at least for life as we know it. No doubt these conditions can be met, but the search for such an exoplanet will surely be protracted. We will need many targets to improve the odds of finding one with life signatures.

The chances that our next generation of telescopes in space will actually discover extraterrestrial life signatures are very slim. The planned surveys are too limited, and the telescopes are too small. We need telescopes with much larger apertures. To detect biomarkers on Earth-like exoplanets will require a more ambitious approach than is currently planned for stand-alone space telescopes. Aperture size really counts, and lunar telescopes are inevitably the giants of the future.

The Moon Is Our Platform

What would be the basic requirements for a future telescope to succeed in the search for life? It all boils down to size. Suppose such a mission happens to not detect *any* biological signatures. We need to be able to at least place a meaningful constraint on the rarity of extrasolar life. Unfortunately, simulations suggest that it may be difficult to make any definitive statement about what nondetection means. The numbers are too sparse, and many unexplored avenues for possible signals remain. A larger sample helps.

One useful goal is to take a range of the strongest signatures of biological activity. An adequate sample of targets in the habitable zones around Sun-like stars would be large enough for us to see whether there is any evidence for biological signatures even if the phenomenon is rare. Only a small fraction of the targeted exoplanets, perhaps 1 percent or even less, might have suitable conditions for the emergence of interesting signatures. Therefore, any prospect of success requires that we have the ability to image and characterize the atmospheres of, ideally, many thousands of Earth-like candidates.

An ambitiously large space telescope capable of searching in the ultraviolet, optical, and infrared spectral bands is planned for the next decade. This 10-meter-class telescope would be a natural mission candidate to search for the simplest indications of life. We could set interesting constraints on the rarity of life out to about 100 light-years from the Earth. But an aperture of 10 meters may simply not be enough to detect a meaningful number of indications of life in nearby exoplanets. To perform a significant search for life, we need to go much deeper in our search, to thousands of light-years.

The problem is that any larger telescope is prohibitively expensive as a stand-alone project. Our real limitation is the size of such a proposed space telescope. This limitation in large part is budgetary. How do we scale up our efforts? How do we detect the billions of Earth twins in our galaxy? We need very large telescopes in spacelike conditions. Powerful telescopes are needed to mount a significant search for actual signatures of life. Size counts.

The real future for Earth-like planet searches lies in constructing telescopes on the Moon. Only there could we hope to build a sufficiently large telescope. Size will not pose an obstacle on the Moon, nor will budgetary concerns be a strong limitation. Although everything that we plan for the Moon is hugely expensive, the required lunar infrastructure will be developed for other goals as well. A lunar observatory will piggyback on the planned development of the Moon. Building it will be a small incremental addition comparable to constructing another lunar tourist hotel complex.

The Moon beckons. Optical and infrared telescopes in dark lunar polar craters could be designed for goals unattainable on Earth or in space. The Moon has spacelike conditions, immense stable platforms in permanently dark and cold craters, and above all sheer size. The timing is right to synchronize a large lunar telescope with the lunar exploration program. A 300-meter aperture telescope would revolutionize our searches for exoplanet biomarkers, bringing good statistical significance to the data we collect on the biological signatures of huge numbers of habitable-zone rocky planets.

The volume of space we survey makes all the difference. A very large lunar telescope is truly a game changer. Instead of surveying a handful of planets, a sample so small that no epidemiologist would ever consider this size appropriate for a statistical sample of a rare disease, a lunar telescope opens up the horizons. We can now peer deeply into our surroundings, from 100 light-years to a depth of thousands of light-years. The target volume for a 300-meter telescope is billions of cubic light-years. This means that up to one million habitable exoplanet targets would now be accessible. Even if life elsewhere in the Universe is relatively rare, with such a large sample, we might have a real chance of detecting definitive biological life signatures. We need unprecedented telescope size to optimize our searches for life in the Universe.

To work in the infrared spectral region, we could use one of the deep craters near one of the lunar poles. These craters are permanently dark and permanently cold because of their high rims and the limited solar elevation from the poles. Imagine constructing a parabolic telescope whose aperture, at 400 meters, is ten times the aperture of the largest terrestrial telescope. Compared to any conceivable space telescope planned over the next fifty years, this giant lunar scope would have thirty times the resolving power and a thousand times the light-collecting area. We could survey a realistic number of exoplanets needed to search for robust tracers of biological activity on remote Earth-like planets.

Ubiquity

On Earth, our only example so far, it took three billion years for the most basic multicellular life forms to appear. It took another billion and a half years to get to where we are today. A series of contingencies affected this progress, from plate tectonics to asteroid impacts. Somehow we survived and evolved. We are a species capable of space travel and rudimentary interstellar communication through radio reception and transmission.

We do not know to what extent those timescales reveal any meaningful constraints on the emergence of complex and intelligent life. Are we at a midway point in emerging complexity? Or are we approaching the end of civilization? Does an existential catastrophe await us? Perhaps evolution could have speeded up in more favorable environments. Or maybe it was all a giant fluke. There would not be time enough for any competition to arise.

Nevertheless, our present place in the cosmos, on a rock orbiting at a safe distance from the Sun, does demonstrate that something dramatic happened at least once. A challenge for astronomers is to establish whether planetary systems that are older than the solar system and capable of maintaining a biosphere are common in the Milky Way. Realistically, near-future searches for extrasolar life will concentrate on nearby stars.

There are tens of billions of galaxies in the observable Universe. Searching for the life that may be lurking in them is a goal for the long-term future. The timescales are dramatic. The current age of the solar system is about half the age of our galaxy's disk and half of the Sun's predicted lifetime. We therefore might expect that roughly half of the stars in our galactic disk are older than the Sun.

This figure by itself is insufficient to judge how common old, biosphere-capable planets are. We need to consider the predicted life span of the biosphere, not just the lifetime of the host star. Earth's biosphere will survive for another billion years or so. Then our planet will lose all its water within an additional billion years. Both effects will be due to the increasing luminosity of the evolving Sun.

Life-hosting planets would still be dependent on geophysical lifetimes, because they must be capable of geochemical cycling. For example, the carbon cycle—the movement of carbon from the atmosphere to land and oceans and back—plays a large part in determining a planet's albedo (reflected light) and temperature. Hence the carbon cycle controls and self-regulates habitability. Our current carbon-burning frenzy threatens this stability.

Intriguingly, a recent examination of cosmic planet formation history concluded that the solar system formed close to the median age for existing planets in the Milky Way. Most stars in our galaxy had a head start on the Sun. Consequently, about 80 percent of the currently existing Earth-like planets may have already formed at the time of the Earth's formation. If so, most of them have a head start on the Earth of billions of years. That raises a perturbing issue. We may be minnows in a cosmic sea, but we may be the only surviving minnows.

Beyond Atmospheres

Why not go for more than atmospheric signatures? Ultra-high-resolution imaging is the holy grail. If we aim to see surface features, oceans, seasonal changes, and more, we will need to do far more than image a pale blue dot. Imagine a really large telescope in the ideal telescope site: a lunar crater near one of the poles, where it is always dark and cold. Imaging from there is the ultimate goal.

Here is how the ultimate imaging resolution might be accomplished. We can make a really large telescope by coherently combining a larger number of smaller ones. This design does not work too well for optical telescopes on the Earth because of diffraction by the Earth's atmosphere. But the atmosphereless Moon looks appealing as a site for this technology.

French astronomer Antoine Labeyrie has proposed filling an entire dark crater with an array of mounted five-meter infrared mirrors to develop a parabolic bowl configuration. This would focus light at a camera strung by cables high above the crater bed. We could build a lunar hyperscope, an infrared telescope with an aperture of 10 kilometers. The hyperscope would give a direct and detailed glimpse of our nearest exoplanet neighbors. The resolving power of a 10-kilometer aperture hyperscope is 10 micro-arcseconds. This is a huge step beyond anything we can currently imagine. Such superb angular resolution could resolve surface features on exoplanets within 100 light-years from us. We could image oceans and continents that are 1,000 kilometers across.

We could study exoplanets of Earth mass in incredible detail as they orbit the red or yellow dwarfs nearest to Earth. We could resolve geological features and atmospheric clouds, or even liquid water ocean properties, via reflected light. We could resolve cloud structures and lakes, mountains and forests. Most dramatically, we could hope to detect nighttime glows from any extraterrestrial cities on nearby exoplanets. If they exist, such cities would be hard to hide. The prospects are stunning.

A kilometer-aperture lunar crater telescope will provide imaging capability that will address a question about the Universe that fascinates all of humanity: Are we alone? Hyperscopes provide the perfect platform for searching for signals that might shed light on the most dramatic alien signatures of all—ones indicative of advanced technology. Many remote exoplanets have benefited from a head start of millions or even billions of years over the Earth. Potential clues to alien life may be omnipresent. Or of course, they may not. We will not know until we look.

The number of nearby targets that we can view at high angular resolution is small. There may be more promise in another approach that involves looking for radio wave signatures. Such a signature of an advanced civilization might be as simple as a television broadcast. Or it could be a far more sophisticated indicator of the existence of advanced civilizations. Clearly, such signatures are highly speculative, as we have no idea of what to expect.

We have already searched, unsuccessfully, for potential radio signals from alien civilizations. Perhaps such signals are our best bet for detecting advanced forms of life. We could see them across the entire galaxy. Searches for radio signals have piggybacked on routine astronomical programs. The great 300-meter Arecibo radio telescope was built in 1963 in Puerto Rico in a depression in the rock. The quasi-spherical dish surface was in a natural sinkhole, and a receiver was suspended 150 meters above the dish. The aperture was fully filled with adjustable metal plates that constituted an active surface of a giant spherical reflector. Arecibo's searches for extraterrestrial intelligence continued over decades, but the telescope collapsed in 2020, no alien radio signals ever having been found. The search continues with its even larger half-kilometer diameter Chinese successor, which was operational in 2020.

Telescopes on the Moon will take this concept a step further. Radio search techniques can be applied on the far side, where radio interference from the Earth is blocked. The Moon turns on its axis over one lunar month. Lunar spin is tidally locked to its twenty-eight-day orbital time. We always see the same face from the Earth.

We would want to build a giant radio telescope on the lunar far side to limit radio noise from the Earth. It would achieve greater sensitivity than what is attainable from terrestrial telescopes, especially at low radio frequencies. Of course, the main goal of such a lunar telescope would be to probe the dark ages, the last frontier in cosmology. We have seen that it would be an essential complement to a much larger radio interferometer by its ability to resolve out spurious signals. But there is a huge bonus: a compelling search for extraterrestrial intelligence signatures could be mounted with little overhead. Only a completely filled aperture would have the sensitivity to pick up the rare flickers of radio noise that might be clues to advanced technology.

A Proactive Approach: Let's Go There

The nearest planets are about four light-years away, around the star Proxima Centauri. This is close enough that we can imagine sending spacecraft there. We would need to have a spacecraft capable of traveling at one-tenth of the speed of light. This is technically feasible for propulsion by a light sail, which works through the effect of intense laser beams of light on what is effectively a mylar sheet attached to an interstellar spacecraft. The light could be provided from a bank of lasers on the Earth. The continuous pressure of light increases the velocity of the spacecraft to near light speed.

The scheme has been studied and is feasible for a sufficiently tiny spacecraft. Indeed, a project to go to the Alpha Centauri star system and image its planets is currently under study. Project Starshot is a privately funded research initiative by venture capitalist Yuri Milner and Facebook founder Mark Zuckerberg. The goal is to launch miniaturized interstellar probes. The typical mass of the device proposed by Project Breakthrough is a few grams, in the form of a sophisticated electronic microcamera. This device is implanted in a gram-scale silicon wafer and propelled by a meter-sized light sail.

Starshot envisages sending a flotilla of 1,000-centimeter-scale spacecraft, each with a light sail powered by a huge bank of terrestrial or lunar 100-gigawatt lasers. The micro-cameras will attain a velocity of one-tenth of the speed of light and arrive at Alpha Centauri after a thirty-year journey. Each microspacecraft will take only minutes to travel by the remote planet and send images back to Earth. Four years after flyby, we might receive high-resolution images of the nearest exoplanets.

We would be able to resolve surface features and oceans and catch a fleeting glimpse of the closest Earth-like planet. Unfortunately, by then it will be too late to do any follow-up imaging. The Starshot camera flotilla would pass by its target at one-tenth of light speed, never to return.

We have seen that there is an alternative exoplanet imaging strategy: building a hyperscope on the Moon. This should be feasible over a similar timescale to the Starshot project. A hyperscope in a permanently shadowed lunar crater could image remote planets out to a distance of hundreds of light-years from us. This is the only way to get more than a tantalizing glimpse of our nearest neighbors.

Because the answer to the question "Are we alone?" may affect nothing less than our claim to being special in the cosmos, its importance cannot be overemphasized. In any case, we shall never know unless we search, and we have seen that the Moon provides the ultimate platform for continuing that search.

CHAPTER 8

Survival

Human progress is exponential. . . . Computers doubled in
speed every two years in the 1950s and 1960s, and are now
doubling in speed every twelve months. By the end of this
century, the nonbiological portion of our intelligence will be
trillions of trillions of times more powerful than unaided
human intelligence. One cubic inch of nanotube circuitry,
once fully developed, would be up to one hundred million
times more powerful than the human brain. . . . Within several
decades information-based technologies will encompass all
human knowledge and proficiency, ultimately including the
pattern-recognition powers, problem-solving skills, and
emotional and moral intelligence of the human brain itself.

—RAY KURZWEIL, *THE SINGULARITY IS NEAR:*
WHEN HUMANS TRANSCEND BIOLOGY

"Space-ship Earth" is hurtling through space. Its passengers
are anxious and fractious. Their life-support system is
vulnerable to disruption and breakdowns. But there is too
little planning, too little horizon-scanning, too little awareness
of long-term risks.

—MARTIN REES, *ON THE FUTURE:*
PROSPECTS FOR HUMANITY

Will Civilizations Survive?

There are strong reasons to believe that the long-term survival of life on the Earth is under threat. Impacts from human activity provide one example that we are able to control, at least in principle. We might irreversibly pollute the planet, or even blow it apart, with thermonuclear devices. Massive epidemics of plaguelike diseases could become uncontrollable calamities. An asteroid impact could devastate the Earth, although we can hope to take preventive measures by detecting and monitoring the orbits of potential killer asteroids at large enough distances. Other risks are more easily managed, if the will is there. Survival risks must be better understood if we are to evaluate the odds for success of a future search for advanced extraterrestrial biological signatures. We will need this assessment to motivate a lunar telescope program.

In the longer term, the Sun will evolve into a red giant star and swell out to 100 times the orbit of the earth. This will occur in about four billion years. Billions of years earlier, the Earth's oceans will have evaporated. The Earth will burn to a crisp, after losing its atmosphere and its oceans. By then, humanity, or whatever remains of it, should have found safer havens than the inner solar system.

Existential risk may be defined as a situation leading to an outcome that would either annihilate Earth-originating intelligent life or permanently and drastically curtail its potential. Here are some examples of existential risk. There will inevitably be a devastating asteroid impact. Humanity may trigger a nuclear holocaust. Nanotechnology may be misused with irreversible environmental impact. Artificial intelligence may take control. Genetic engineering could run amok. And of course, events unforeseen should be added to this list.[1]

The potential risks are chilling. Our aim here is to search for traces of life on distant planets. The critical issue in that search is rarity: How far do we need to search to achieve a minimal rate of detection of biologically advanced phenomena? We must evaluate the existential risks in order to guide our choice of future lunar telescopes.

When we consider the prospects for humanity over the much longer term, we should question the notion of permanent setbacks to intelligent life. Nature abounds with stories of life surviving through the most extreme catastrophes. True, what survives may be severely limited. But over the course of eons of time, the irresistible drive of evolution is likely to come up with new and unforeseen achievements that are capable of surpassing all previous histories. There is a bright side.

Asteroid Impacts

The lunar surface has been largely shaped by asteroid impacts and associated volcanic activity. The Earth bears witness to a similar history. It is harder to see than on the Moon, as weathering, tectonic activity, and water flows on the land have largely erased any ancient fingerprints. To probe the Moon's impact history, we can use major geological features on the Earth to provide a lunar template. Relics of asteroid impacts on the Earth inform us about similar features on the Moon that would have occurred and that have survived from far more ancient events. Eventually we will have in-site geological surveys of the lunar surface to refine our first estimates.

The best-known terrestrial event was the impact of a massive asteroid of some tens of kilometers in diameter. About 66 million years ago, a massive impact created the 300-kilometer-wide Chicxulub crater off the Yucatan coast. This had global impact and devastated the terrestrial environment. The risk of another such impact tomorrow is about the same as it is for any traveler dying in a commercial plane crash. The risk is not completely negligible.

The impact that created the Chicxulub crater probably led to a short-lived but devastating phase of global climate disruption. Dense atmospheric layers of incandescent dust, ash, soot, and other compounds were generated. The impact had knock-on effects on the Earth's cloud cover that led to a severe extended winter. These events are likely to have triggered mass extinctions of many species. Photosynthesis in plants was halted, and plankton largely died out. Some 75 percent of the plant and animal species on the Earth were abruptly destroyed. The geologic record reveals a thin layer of iridium-rich sediment at this point in time. Iridium is rare on the Earth but common in asteroids.

Volcanic events may also have been triggered by the impact.[2] These amplified the severity of the sudden asteroid-triggered climatic change. All of this happened simultaneously over a relatively short period of time, as expected for an impact event. Our best estimate is that the impact came from an asteroid some 15 kilometers across. Imagine what the impact of a 100- or 1,000-kilometer-size asteroid might do.

There were certainly some benefits of the extinction event. Its effects were not all negative in that it created some new evolutionary opportunities. Dinosaurs did not survive, and their extinction may have opened a pathway for the predecessors of *Homo sapiens*. Large mammals such as horses and whales, as well as bats, birds, and fish, proliferated in newly available ecological niches. Nevertheless, the outcome of that early asteroid impact could have been so much worse.

Recognizing that it would be good to have advance warning of a potentially disastrous asteroid impact, we are searching for orbital indicators of future asteroid collisions with the Earth. Our current focus is on the near-Earth asteroid Bennu, which has a diameter of half a kilometer and would have 100 times the explosive power of the Tunguska event. It would most likely have a devastating impact if it hit the Earth, although its unknown composition introduces uncertainty in predicting the impact effects. The OSIRIS-REx spacecraft will return asteroid rock samples to Earth in 2023.

Bennu orbits the Sun. For those looking for a precise date of a possible future impact, its next-closest approach to Earth is predicted for September 24, 2182, when it will come within 460,000 miles. This is dangerously close. This prediction is based on estimates of closest approach, however, that are uncertain. All we can say now is that the odds are one in 2,000 for Bennu to impact the Earth by 2300.

On average, asteroids the size of Bennu are expected to impact the Earth once every 100,000 years. These would be rare events that could trigger widespread devastation. But they are not serious existential threats to terrestrial life.

There are much rarer and far more catastrophic events that occurred over timescales spanning hundreds or thousands of millions of years in the past. The early Earth suffered many meteorite impacts. We infer this from studying the Moon, where the traces of earlier impacts have survived. Surface erosion has been largely ineffective at deleting traces of old craters on the lunar surface. The lesson from the Moon is that meteorite impacts occurred over several billions of years, and that the impact rate was greatly accelerated over the first billion years of lunar history.

Sooner or later, a truly major impact will occur again on the Earth. It would be hundreds of times more powerful than the dinosaur-destroying hit that created the Chicxulub crater. This would be bad news for humanity. We need to keep a constant vigil. No doubt humanity would survive, but it is likely that civilization would be greatly diminished. It is somewhat of a random chance in a cosmic lottery.

Existential Risks

There are many risks awaiting humanity that could severely reduce the lifetime of any civilization and hence its potential detectability. Let's assess the most worrying of these.

Global Warming

A large part of global warming is attributed to pollution by aerosols and carbon dioxide.[3] Methane and water vapor also play important roles. These gases absorb sunlight, which is in turn converted into infrared radiation. One consequence is the greenhouse effect, in which the infrared light is trapped by the atmosphere. This results in global warming. The torrid atmosphere of Venus is an example of a runaway greenhouse effect.

The most alarming aspect of global warming will be the rise of the ocean level. It is estimated that at most 10 percent of the world's population will be affected in the most vulnerable countries. Even including the risks of social unrest and mass immigration, is this enough to cause massive global upheaval?

Atmospheric pollutants are a man-made contribution to the greenhouse effect. So is the carbon dioxide generated by the burning of fossil fuels. We can monitor the rise in atmospheric temperature since the industrial revolution of the eighteenth century. That epoch was marked by the onset of a massive phase of coal burning. The climate subsequently changed as the carbon dioxide concentration increased by 50 percent. The mean temperature has risen by 1 degree Centigrade since the beginning

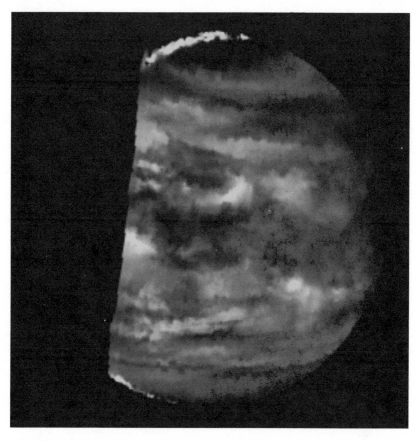

FIGURE 8. Sulfuric acid clouds on Venus and toxic rain. The European Space Agency's Venus Express studied the planet from orbit between 2006 and 2014. This image, in ultraviolet light, of the clouds of Venus shows the 20-kilometer-deep cloud layer, known to be mostly composed of tiny droplets of sulfuric acid that rain on the surface. The thick atmosphere traps solar heat radiated by the surface, leading to a runaway greenhouse effect.

Image credit: ESA © 2007 MPS/DLR-PF/IDA.

of the industrial revolution around 1750. This seems like a small rise, but it was enough to raise the ocean level through glacier melting by 23 centimeters over the past 200 years. This is just a foretaste of what awaits us. Many people live at or near sea level. Perhaps hundreds of millions of people would be displaced by further ocean level rise.

Other natural catastrophes are thought to be associated with global warming. There may be an enhanced incidence of seismic and volcanic activity. The consequences of fossil fuel burning add to the mix. Global warming is inexorably boosted by the known effects of burning coal over many decades. A major side effect is acid rain. This is something that has affected many areas of the world where coal is utilized as a fuel. There are disastrous consequences for forests and agriculture. Even if we succeed in controlling acid rain, global warming is something we may have to live with.

Thermonuclear Holocaust

From the earliest studies of nuclear bombs, scientists worried about the possibility of triggering a global nuclear catastrophe.[4] These considerations first arose in the 1940s, when the concept of a nuclear chain reaction was first realized. Could there be a similar chain reaction on a global scale as more and more powerful bombs were tested?

These warnings did not deter the drive for detonations. Since nuclear physicists' calculations had been reassuring, atmospheric testing of nuclear weapons proceeded throughout the 1950s, culminating in the Russian Czar hydrogen bomb, a 50-megaton monster that exploded in the atmosphere in 1961. This was 1,000 times more powerful than the atomic bomb used at Hiroshima. The devastation from nuclear accidents and the health risks from atmospheric testing all played key roles in the negotiations that led to the nuclear test ban treaty in 1963.

The testing of nuclear weapons is no longer considered to be a major risk factor. Triggering nuclear war is another matter. There have been incidents that led to close calls in the past half-century. Most notable was the high tension in 1962 of the Cuban missile crisis. Strategic Soviet nuclear missiles were installed within seventy miles of the US mainland. President John Kennedy himself estimated at least a 30 percent chance that the clash with Khrushchev and Castro would end with nuclear war.

Fortunately, cooler heads prevailed. But it is worth speculating on what might have happened. Our best guess is that a local nuclear war in 1962 would have had tens of millions of casualties. The possible deployment of the vast stockpiled nuclear arsenals poses an even greater risk, and one that is hard to assess. Today there are new players on the nuclear scene. Worldwide nuclear conflict seems unlikely, but the consequences could still be devastating.

The most extreme risk scenarios arising from nuclear conflict involve not so much radioactive contamination as the onset of a nuclear winter. The fires and smoke from a nuclear holocaust would uplift into the stratosphere. Sunlight would be blocked for some five years according to atmospheric circulation and climate modeling. The average global temperature would drop by about 7 degrees Centigrade. This temperature drop would have an immense impact on agriculture, especially in temperate areas. The reduction in food production would trigger mass famines that would set back civilization by decades or more. But humankind is sufficiently robust to survive. Perhaps lessons would be learned by future generations.

Genetic Engineering

In the wrong hands, genetic engineering could be a biohazard. Mass diseases and plagues historically have presented a natural risk for humanity. The bubonic plagues and flu epidemics are examples in Europe. Most recently the Covid pandemic and its various strains have caused millions of deaths. But again, human ingenuity is capable of coping with such tragedies and of ameliorating future risks via a worldwide vaccination program.

Such hazards are fed and amplified by biotechnology, in which radical advances are made that are difficult to monitor and control. Somewhere a pharmaceutical start-up will place profits ahead of ethics and fail to contain a self-replicating pathogen. According to futurologist Ray Kurzweil, "a self-replicating pathogen, whether biological or nanotechnology based, could destroy our civilization in a matter of days or weeks."

The rapid advances in designer gene technology have raised concerns that a terrorist plot or even a mischievous adolescent could design molecules with fatal consequences for humanity. An industrial accident could equally lead to the inadvertent spread of fatal viruses. Historically there have been several accidental releases of dangerous bacteria or viruses. Fortunately, these were soon controlled. But assessing future risk from such biohazards is highly uncertain, and the consequences are unpredictable.

Other risks include the impact of major population shifts and changes in lifestyle. Interaction between the modern world and historically isolated communities has transferred diseases to the latter, with devastating effects. This is how indigenous life in the Americas and in Australia has been effectively destroyed. Perhaps this is just Darwinian evolution at work. *Homo sapiens* prevailed over Neanderthals some 200,000 years ago, after a slow transition from what may have been coexistence for over 100,000 years. But the outcome was inevitable. Brain capacity most likely played a key role in humanity's adaption to diverse environmental pressures. Still, one feels anguish for the losers.

The human species is certainly still evolving, and on a timescale that, however slow, will still be immensely shorter than the remaining lifetime of the Sun. We cannot foresee where natural genetic evolution may take us by then. Certainly genetic engineers will be tempted to speed up the process. Perhaps humanity could become immortal.

How to control a quest for immortality is unclear, especially as the advantages will be designed to outweigh any negative aspects. This is not a path to the future that seems morally desirable, in part because it allows a small minority of scientists and politicians to irreversibly control that future. The outcome would leave unlimited power in the hands of our leaders. This is not an ideal solution. History tells us that even democratically elected leaders have on rare occasions misused their powers. Let us hope that humanity will be able to retain control over such ethically bleak prospects for the future.

Tiny Black Holes

Physics hazards exist.[5] Particle colliders smash protons together at very high energies when physicists are trying to unlock the inner workings of protons. Collider experiments led to the discovery of quarks, particles that exist when confined by nuclear forces within atomic nuclei. Quarks are the building blocks of protons and neutrons, and along with electrons and neutrinos they constitute the basic elements of the standard model of particle physics.

We are confident about this model. It predicted the existence of the Higgs particle, detected in 2012, that explains how other particles acquire mass. However, the model leaves us with a huge gap to fill. The quantum theory tells us that when we include gravity, the maximum mass of an elementary particle is about one-hundredth of a milligram, which is about 10 million trillion proton masses. This is the Planck mass, which is achievable soon after the onset of the Big Bang, thanks to the huge energy available at this brief instant of time. Of course, we have no direct access to this epoch. How can we test this ultimate limit to particle masses?

Collider experiments will not get us anywhere near this limit. Current energies at the world's largest particle collider, the Large Hadron Collider (LHC) at CERN (European Organization for Nuclear Research), amount to the energy equivalent of more than 10,000 proton masses. To test the quantum theory, we would need to achieve the much higher energy of the Planck scale. That's 1,000 trillion times higher. In between there is a particle desert—or perhaps a paradise teeming with new but evanescent particles. We have no idea. The larger the particle accelerator, the higher the energy that is achievable as we smash particles together. To arrive at the highest imaginable energy scale, we would need a collider stretching tens of billions of times the distance between the Earth and the Moon. That is, 1,000 light-years. This is impractical.

There is certainly a small risk involved as we do build ever more powerful particle colliders. Particle collisions at high enough energy could create tiny black holes, as predicted by theories of quantum gravity. The theories also tell us that the energies required are vastly higher than any terrestrial particle collider could ever achieve. And there are theories about where microscopic black holes can be made in higher dimensions and then unleashed by high-energy particle collisions. Most likely they evaporate immediately. But how reliable are the theories? Without experimental verification, we do not really know.

One concern expressed by some scientists is the idea that if there are enough collisions in a terrestrial experiment, one exceedingly rare event might attain the quantum limit. It would be just a fluctuation. And fluctuations happen, albeit rarely. Normally we could routinely reproduce such extreme and potentially black hole–producing energies only in a particle collider vastly larger than the Large Hadron Collider. But rare ultra-energetic events can occur. If our quantum conjecture is correct, a black hole could be created. Production of even a tiny black hole could cause a disastrous implosion. Matter would rush in and the black hole would grow. The entire Earth would be at risk.

Nuclear physicists carefully studied this possibility during the construction of the Large Hadron Collider. Their goal, required by a court decision, was to produce an environmental impact statement. Theorists had earlier speculated that higher dimensions are present on scales smaller than the Planck scale. This is still equivalent to energy scales far higher than those accessible by current particle colliders. But the new machine might take us closer to the precipice. Unlocking higher dimensions would permit the formation of tiny black holes.

The researchers concluded that such events should already have occurred at the far higher energies naturally achieved as cosmic rays from outer space bombard the Earth's atmosphere. No such creation of tiny black holes was ever seen. Hence the risk is low. Construction design for the Large Hadron Collider succeeded in passing this environmental risk assessment.

Nanotechnology Run Amok

Atomic-scale nanotechnology is a new development in manufacturing that can deliver highly miniaturized factories. It is
feasible that robots no larger than bacteria could produce industrial products. Assemblies of such nanobots would generate
copies much as a 3-D printer does, but with molecular precision. So far this is just a dream, but atomic-scale nanotechnology is eminently feasible in the near future.

This seems to be an attractive use of future technology. The
risk is that there would inevitably be an irresistible commercial
drive to build self-replicating machines. These would greatly
increase production efficiency in manufacturing facilities. In
principle, the only limit would come from the supply of raw
materials. Warnings have been voiced about the misuse of such
technology in creating nanobot factories that could wreak
havoc with the economy and especially with the environment.
Systems built to self-replicate could have a wide environmental
impact.

Other consequences would be even less well controlled. Unlimited supplies of the most powerful weapons could be manu-
factured. Accidental or even deliberately engineered mutations
might be equally destructive and potentially massacre the biosphere. An extreme vision of self-replicating nanotechnology
would see the runaway replication of nanobot manufacturing.
Unlimited self-replication could create havoc. All raw materials
would be consumed, and not much else would remain un-
used in this most efficient process. This view of the future has
been pessimistically characterized as the dominance of "gray
goo." Some habitable planets might have attained this regime,
with corresponding implications for their potential biological signatures.

Of course, warnings about potentially bad side effects will not be sufficiently persuasive to halt nanotechnology research. Although current nanotechnology applications provide enormous societal and commercial benefits, the future of nanotechnology may be hard to control. Existential risks to our future are a concern for advances in nanotechnology as with other new technologies. International organizations must be called upon to develop a nanotechnology safeguard system, taking an active shield approach. This would allow us to harvest nanotechnology's positive aspects while controlling its adverse implications.

Artificial Intelligence

Computers of unprecedented power could render humanity irrelevant. The rise of the machines has been touted as a possible future. This is given the enormous advantages, both physically but especially intellectually, that robots would acquire over anything feasible for humanity. We already are at a stage where in the world's most sophisticated game, Go, computers routinely defeat the world's champion players.

Flashback to 2017. A new computer program including neural network learning, called AlphaGo, had just been generated by a Google subsidiary, artificial intelligence developer Deep-Mind, based in London. The superiority of a similar program over the world's best chess players had already been demonstrated. Next up, the game of Go, far more complex than chess, was the real challenge. AlphaGo, and its successor AlphaZero, learnt how to achieve superiority over the world's leading Go players. This happened in less than 24 hours of training and playing against itself. Go master Fan Hui, was the first professional Go player to lose to AlphaGo. He commented about the winning move over the world professional Go champion Ke Gie: *Its not a human move. I have never seen a human play this move. So beautiful.*

Artificial intelligence will soon be capable of exceeding the intelligence capacity of the human brain in regimes far beyond gaming. There is a debate about how long this will take. Estimates by experts give a significant chance, say 50 percent, for this to happen by between 2050 and 2090.

One concern is that advanced computers would be capable of making self-replicating artificial intelligence machines that improve in each successive generation thanks to deep learning techniques. Whether they would be aligned with human values is a debatable issue. Artificial intelligence could potentially be malevolent towards life. Humans might be regarded as mere irritants, a source of primitive interference. Humanity might be relegated to a zoo.

Why couldn't we just turn off the switch and reboot? This is fine when your laptop malfunctions. But an advanced AI system could easily circumvent such human control. It could bury its machine code in countless digital storage sites. Once in the data cloud, it might be difficult to isolate and suppress. We already face similar issues with computer viruses. This would be the merest foretaste of what awaits us in a world dominated by artificial intelligence machines.

We are on the verge of developing quantum computers. These tap virtual numbers. As a consequence, their power is potentially infinite. Non-biological brains have no limits. Whether with sufficient computing power, they can develop self-awareness is less clear. Something akin to emergence of consciousness is debatable. However given how far in complexity an artificial brain could evolve relative to the neuron-firing limitations of a human brain, it is difficult to see any endpoint.

The issue then arises as to whether such inevitable progress in artificial intelligence would have positive consequences for humanity. One can certainly program efficiency. That is undoubtedly beneficial. It makes us all wealthier. Programming wisdom or compassion may require higher dimensions of neural networks and self-learning than we can currently imagine.

One of the more bizarre predictions that arises as a consequence of unlimited computer power is that in some remote future, should humanity survive, our descendants will have developed such powerful computer games and simulations that they could reproduce the history of humanity in detail. Philosopher Nick Bostrom argues that they would have created the ultimate computer game. The simulation would be indistinguishable from the real thing One could explore the contrafactual evolution of humanity and of the cosmos. There would be parallel histories. We would live in one. The existential risk is that someone from the future could pull the plug, and we are extinguished. But there must be hope that there are ways to safeguard against such attempts at sabotage.

Next Steps for Humanity

Assessing existential risk has become something of an industry in recent years. Gloom prevails in risk assessments made on timescales as short as centuries. Some of the risks that lie ahead are capable of driving dangerous life-threatening mutations. Our recent experience does not give high confidence in a favorable outcome. Pandemics from mutated viruses, for instance, have had negative impacts on civilization. But humanity has recovered, wiser, sadder, and stronger for the experience, whether it was the bubonic plague, influenza epidemics, or the Covid crisis.

We need a balancing act, but that may be beyond our control. We can blame the environment, the atmosphere, and natural disasters from megavolcanos to asteroid impacts for potentially catastrophic changes. Survival is a delicate process, but optimism has a role too. Some events might even impact our genetic code in a positive way. The long-term view provides a more positive spin, a chance to declare the glass half full, not half empty. Human evolution is controlled by genetic mutations. Without a past of vigorous mutations at the dawn of life, we would not be here now. Perhaps existential crises lead to a positive outcome over the course of billions of years. Those are the evolutionary timescales we must contemplate as we survey the stars.

We need to plan for the long-term future. This is difficult as most governments have a planning horizon of less than a few years. Administrations in democratic nations are very much dependent on election prospects. Nevertheless, we need to develop a longer-term perspective, especially in science. Exploration of the Moon, most likely our next significant step into space, demands such a vision. So for that matter does building the next generation of particle colliders, which will necessarily be at least ten times more powerful than the Large Hadron Collider. Beyond science goals, we need to better assess the risks that lie ahead over the next centuries, at the very least.

Space can be a dangerous place. Someday we will establish space colonies. Careful planning should enable us to avoid many of the mistakes we have made on planet Earth. Perhaps the first global barrier to surmount will be that of space debris. Low-Earth orbit, where many thousands of pieces of rocket launchers and satellites orbit at speeds of some 10,000 miles per hour, is an especially crowded environment. As many as one million fragments of metal or plastic larger than a centimeter are in low-Earth orbit. Impact with a spacecraft could have dangerous consequences. Fortunately, most of this larger debris is being tracked, although the amount of debris is increasing rapidly.

The Long-Term View

The Universe will eventually exhaust its energy supply, even though stars are generous donors. They return about one-third of the gas they consume to the interstellar medium. It will take a while to run down the gas supply, but once new stars can no longer form, it will be all downhill, like an aging society. There will still be planets, surrounding white dwarfs and neutron stars, and there will even be black holes. As planets orbit their compact hosts, tidal drag forces will heat their interiors. The ultimate chill will be avoided. Life will survive even in such cold environments.

There are prospects for extracting electromagnetic and gravitational energy from compact stars. Gamma ray bursts, the most luminous phenomena in the Universe, periodically release energy from highly magnetized neutron stars. Ultraluminous X-ray sources are generated as black holes accrete matter from close companion stars. These are powerful sources of release of gravitational energy. Accretion of gas onto black holes releases energy at high efficiency. The gravitational release of energy can be some ten times more efficient than energy released by the fusion of hydrogen in stars. Capturing this energy source is our best bet for achieving immortality—or getting as close to it as is feasible.

Of course we cannot live too near a black hole. We need the comfort of the solid surface of a planet, for example. But planets could happily orbit black holes. There is an ultimate limitation here, in the very distant future. The end approaches when protons will decay. Proton decay is a prediction of the unification of physical laws, something that physicists generally believe must occur at sufficiently high energies. There will no longer be planets or rocks, nor even stars. All of this will happen within about a trillion trillion trillion years from now. That leaves us with black holes.

Even black holes are not forever. As Stephen Hawking discovered, black holes evaporate. In a small black hole, the curvature of space is so high that the very fabric of space self-destructs in a burst of radiation, and the black hole evaporates. The bigger the black hole, the more stable it is. A black hole with the mass of a mountain evaporates over the age of the Universe. We have not actually observed this, and perhaps such tiny black holes never existed. But the theory is generally accepted. It tells us that the massive black holes produced by stellar deaths are much more long-lived. These black holes survive a million times longer than protons, but even they are fated to decay. This remote epoch should mark the end for any remaining semblance of intelligent life. At the very least, life needs a supply of energy.

If we are prepared to speculate more wildly, quantum theory offers a possible solution for survival into the infinite future. The quantum vacuum generates fluctuations that come and go on a timescale too small to be measurable. Disembodied brains can appear out of nothing. No need for bodies. According to quantum cosmologists Andreas Albrecht and Lorenzo Sorbo, "The most likely fluctuation consistent with everything you know is simply your brain (complete with memories) fluctuating briefly out of chaos and then immediately equilibrating back into chaos again."

Perhaps this is too extreme a fate for most followers of this discussion. Here is a novel piece of quantum thinking. There was no Big Bang, at least not from infinite density, but there was a Big Bounce. Particles can act like waves and avoid hard collisions with each other. Their trajectories can take them into otherwise inaccessible regions of space. This is a manifestation of quantum uncertainty. A treatment of the expanding Universe as a quantum entity has led some advocates to argue that the Universe began as a Big Crunch via collapse from some almost infinitely large state. At some maximum density, near the Planck limit, when particles and black holes are indistinguishable, the Universe switched from collapse to expansion. Whence the origin of the Big Bang.

As a theory, there are many loopholes in this reasoning. There is no evidence for such a Big Bounce. But at least philosophically, it answers a fundamental question: Where did we come from? It provides us with an infinitely long period of contraction that preceded the expansion phase. And even the future might be affected if expansion were to peter out.

If there is indeed a cosmological constant, the Universe will accelerate forever. But if the cosmological constant is not truly a constant but just its physical manifestation, the energy of the vacuum may decay. Then another fate is possible. The Universe could reach a maximum size in a few tens of billions of years, then recollapse to a future Big Crunch. And the series of cycles would continue, to an infinite future from an infinite past.

On a more mundane note, one piece of good news is that the near-term future is bright, if we survive our existential crises. We can outlive our Sun. Catastrophic stellar failures, such as the one that the Sun will experience in a few billion years, do not apply on a similar timescale to red M-dwarfs. These stars are cooler and lower in mass than the Sun. Such stars evolve extraordinarily slowly. They and their planets have something like 100 billion years ahead of them. Moving to the vicinity of a nearby M-dwarf is our best bet for the long term. Forecasting the future of the Universe opens vast vistas of time for evolution to achieve its full potential.

Searching for Intelligent Life

The evolution of intelligent life involves many open questions.[6] What are the geochemical constraints on the evolution of complex life? On what timescales do these constraints operate? Is there, as it seems, an impetus toward biological complexity? Are there any evolutionary bottlenecks that make it extremely hard to make the transition to intelligence? Are there existential factors that limit the life span of intelligent species?

We would ideally like to go beyond simple biological signatures and seek the clearest sign of an alien technological civilization. This could be the unambiguous detection of an "intelligence" that might contain non-natural signals. One often-invoked method is via radio transmission or infrared beaming via lasers. The clearest sign of an alien technological civilization could be the unambiguous detection of an information-containing, non-natural signal. So far, however, no such signal has been reported.

To find a sufficient number of candidate planets, we would need to survey many thousands of targets. The implied remoteness puts them hundreds or more likely thousands of light-years away. That sets the most probable timescale for a conversation. Patience will be essential. Responding to the transmission of radio signals from another galactic civilization—assuming one exists—would require 1,000 years or more. It would be a slow discussion.

A recent attempt to estimate the fraction of the galaxy that has been reached by radio communication from the Earth

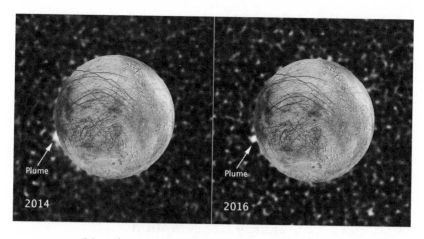

FIGURE 9. Water plumes venting on Jupiter's moon Europa. These composite images, photographed in ultraviolet light by the Hubble Space Telescope's imaging spectrograph in 2014 and 2016, show suspected plumes of icy material erupting two years apart from the same location on Jupiter's icy moon Europa and rising up to 100 kilometers above Europa's frozen surface. The plumes provide evidence of a global water ocean under a frozen icy crust. There is a remote possibility that life might exist in these oceans. The snapshots of Europa were assembled from data from NASA's Galileo mission to Jupiter.

Image credit: NASA/ESA/W, Sparks (STScI)/USGS Astrogeology Science Center.

concluded that only about 1 percent of the galaxy has hitherto been probed. To give us reasonable odds, we might want to reach something more like 50 percent of suitable planets before a signal can be expected. This puts the probable time for a reception of such a radio signal from another galactic civilization—again, assuming one exists—a few thousand years into the future. But don't hold your breath—the uncertainties are considerable.

We can try to do the calculation in a more sophisticated way. Let's use an argument based on statistical reasoning to evaluate the possibility that life will emerge in abundance around low-mass stars trillions of years from now. Our best estimates are that the distant future holds much more promise for interstellar communication than the current period, because our one example of intelligent life appeared on the cosmic scene rather early. The fact that there is much more time ahead weights the odds.

Radio signals may not be the optimal route. There is a distinct possibility that radio communication would be considered archaic by an advanced life form. Its use might have been short-lived in most civilizations. It would then be rare over large volumes of the galaxy or the Universe. So we need to be cleverer.

Energy Consumption

What might then be a more generic signature?[7] Energy consumption is one hallmark of an advanced civilization that appears to be virtually impossible to conceal. Energy is essential. One of the most plausible, long-term energy sources available to an advanced technology would be a command of stellar luminosity.

The prime source of energy is the stars. The engineering construction needed to use entire stars as energy sources is known as a Dyson sphere. Anglo-American physicist Freeman Dyson suggested that an advanced civilization might harvest the starlight from a star, or even from many stars. This potential long-term energy source would be controlled by thermonuclear fusion of hydrogen in stellar cores. It would provide unlimited resources if we could harness nuclear energy on stellar or galactic scales.

In all cases, even with the expected higher-efficiency energy production of an advanced civilization, waste heat would be an inevitable outcome. One consequence would be the production of a detectable signature in the mid-infrared spectral region.

One concern is that even in the absence of advanced technology, natural emission from a circumstellar dust belt might confuse any putative artificial signal. But this might be distinguishable by spectroscopic observations. It is in the infrared region of the spectrum where our future lunar telescopes will provide unprecedented sensitivity. Infrared emission seems almost unavoidable, and impossible for future civilizations to hide. Note that the anticipated infrared signal of a Dyson sphere should be a non-negligible fraction of the luminosity of a star. It should far exceed typical reflections from terrestrial planetary surfaces and atmospheres.

There may be even more dramatic advanced signatures. Russian astronomer Nikolai Kardashev argued that the most extreme case might involve capturing the starlight from an entire galaxy. Here is the possible signature. The starlight would be converted into infrared radiation by many Dyson spheres. A truly advanced civilization would operate galaxy-wide. In our search for such a civilization we would look for galaxies that are bright in the middle of the infrared range and, at the same time, deficient in starlight at shorter wavelengths. There is no simple astronomical explanation for such an energy distribution. Such systems could be proxies for efficient users of starlight for advanced technological purposes.

A recent large survey by NASA's Wide-Field Infrared Survey Explorer (WISE) satellite did identify a few red spiral galaxies as potential candidates for advanced civilization status. The survey mapped 100,000 galaxies. A handful of oddballs were found that emit an excessive amount of infrared radiation. We know they have a high rate of star formation. They should be teeming with young stars and be blue, but they are red. Something is swallowing most of the starlight, and it is not just interstellar dust clouds—their radiation is too cold. Perhaps we are seeing a galaxy full of Dyson spheres.

A more conservative explanation for these observations would be that unusually large amounts of interstellar dust are present close to the stars. The dust would need to be warmer than typical interstellar dust. Perhaps this could naturally be arranged for a certain type of dust. Artificial artifacts might not be needed. All we can really conclude is that such peculiar objects definitely deserve follow-up observations. We need to explore in much more detail whether there are more natural explanations. Any of these bizarre galaxies might represent the signature of a remote galaxy-dominating civilization. Perhaps there would be unusual radio signals, or other multi-messenger indicators. We need to look.

Technosignatures

Potential signatures of technological civilizations include various forms of atmospheric industrial pollution as well as short-lived radioactive products. These effects are necessarily transitory. Basically, we expect that aliens either learn how to clean up their act or destroy themselves.

More pessimistically, biologically based intelligence may constitute only a very brief phase in the evolution of complexity. This would be followed by what futurists have dubbed the "singularity." This is the moment when artificial, inorganic intelligence inevitably dominates. Many regard the singularity as an unsubstantiated projection. If this is indeed the case, however, advanced species are not likely to be found bound to a planet's surface. Gravity is helpful for the emergence of biological life but is otherwise a liability. Rather, floating freely in space seems to provide a more desirable and controllable environment.

Given this imaginative conjecture about the distant future, we can still argue that, simply because of energy requirements, any surviving species must be near a fuel supply, namely, a star. Detection is less evident. Even if such intelligent machines were to transmit a signal, it would probably be unrecognizable and non-decodable to our relatively primitive, organic brains.

Other transient signals may be more promising. Nuclear war is a possible source of radioactive atmospheric pollution, but presumably a short-lived one, essentially gone after 100,000 years. Industrial pollution lingers for a much longer time, but we may hope for advanced cleanup technology. Fortunately, life is remarkably resilient. Allowing for the worst excesses of nuclear war, global warming, or a major asteroid impact, advanced life should regenerate over tens of thousands of years. This means that if civilization self-destructed after 10,000 years, it would regain and likely surpass its earlier peak after another 10,000 years had elapsed. We should make fewer mistakes the next time around.

We might expect to need rather more than 1,000 years of civilization—roughly the time since the Battle of Hastings—for humanity to approach a climax for the full potential of auto-destruction. Perhaps my estimate of 10,000 years is too generous. We have the power today. Within a few hundred years, our capabilities will have greatly increased. A catastrophe would set civilization back to the level of the Middle Ages or even, in a worst-case scenario, back to the Stone Age. Let's hope we acquire enough wisdom to avoid this dismal fate.

The good news is that life is incredibly resilient. On Earth, it exists in volcanic craters and in the deepest depths of the oceans. We can be more optimistic. According to the one case study we have—ourselves—any anthropic disaster should allow recovery over relatively short periods, perhaps as short as thousands of years. We have already achieved much in a few thousand years. I estimate that a time sufficient for regeneration to a new state of humanity must be on the order of tens of thousands of years. Such a level of civilization should then be capable of the highest accomplishments of philosophy and poetry, of art and science. We would be equally capable of devastating wars and interplanetary travel. There would inevitably be high degrees of existential risk. And maybe the cycle continues.

The fragility of civilization could perhaps explain the Fermi paradox. If such a scenario holds true, our chances of detecting simple life may far exceed those of discovering intelligent extraterrestrials capable of interstellar travel. Simplicity is so much more prevalent because it lasts longer. This reasoning holds even if life is ubiquitous.

Still, detecting the signature of an advanced intelligence, whether biological or nonbiological, remains the most intriguing goal. If we are optimistic about existential risk, advanced life forms will dominate whatever there is to be found. All power to the proposed projects for the 2020s that will explore the infrared regime where some of these more exotic signals may be lurking, such as the proposed Japan-led Space Infrared Telescope for Cosmology and Astrophysics and NASA's proposed Far-Infrared Surveyor satellite telescope.

The key point is that for the first time in human history we are perhaps only a few decades away from being able to actually answer the question, "Are we alone?" This might just be the most important question we have ever posed. The real major advance will come with lunar hyperscopes, which will allow us to finally collect very large samples of habitable exoplanets. The search will be able to proceed with reasonable statistical confidence. Once the first extraterrestrial biological signatures are reported, collateral programs will be strongly enhanced. Even now, there are compelling reasons to take the high-risk road of searching for technological signatures of more advanced habitats. Searches for electromagnetic signals from other galactic civilizations will undoubtedly accelerate.

Have We Been Visited by Aliens?

The ultimate technosignature would be that left by a visiting alien. There is a long history of sightings of unidentified flying objects, popularly viewed as spacecraft shaped like flying saucers. Such claims have traditionally been debunked by scientists who find plausible natural explanations.

Another type of alien signature, some have claimed, can be seen in the form of terrestrial relics of alien visits, such as the Nazca Lines in Peru. These surface features, which are most likely of geological origin, have an unusual degree of regularity. One alternative speculation is that they are markers left by an ancient alien civilization. This story was recounted long ago by Erich von Daniken in the popular best-selling book *Chariots of the Gods*. Again, there are invariably plausible natural explanations. Posing such extreme hypotheses violates scientific credibility when mundane and plausible, if unproven, explanations can account for the phenomenon in question.

Claims of alien visits will not go away. Recent debates about such encounters have been reinvigorated by the discovery in 2017 of an asteroid-like object of interstellar origin. The discovery was made during a routine sky survey searching for potentially dangerous asteroids that might impact the Earth. The survey telescope is located on top of the Haleakala volcanic peak in Hawaii. Named Oumuamua, after the Hawaiian word for "scout," the object has a highly unusual orbit and a highly

unusual shape. From studying the reflected sunlight, astronomers deduced that Oumuamua is moving at high speed away
from the Sun. The light variations tell us that it is highly elongated and tumbling in its orbit.

Normally, we explain the nongravitational acceleration of
comets as due in large part to outgassing from a massive block
of ice. The outgassing produces jetlike emissions that accelerate
the icy rock. Cometary tails are generated as the object plunges
in toward the Sun from the cold depths of the outer solar system,
where comets are born in a vast cloud of orbiting icy bodies. If
the velocity is too high to be explained by solar gravity, most
likely the comet originates from a nearby star. These too have
clouds of comets.

Oumuamua's high speed is hard to understand. There is no
evidence that it has a cometary tail. The failure of an astrophysical explanation for its orbit led Harvard astronomer Avi Loeb
to argue that Oumuamua is an artificial construction of alien
origin: "Oumuamua didn't behave as an interstellar object would
be expected to because it wasn't one. It was the handiwork of
an alien civilization." Few astronomers take this argument
seriously, simply because more conventional astrophysical
explanations are available.

The most reasonable idea is that the natural outgassing that
propelled the comet was mostly over when the object was discovered. Any expected tail would then be too elusive to detect.
The inferred giant icicle-like shape was produced as the block of
ice evaporated from overexposure to solar ultraviolet radiation.
Oumuamua's peculiar trajectory and tumbling may be in part
due to highly asymmetric outgassing, which acts like rocket engine
exhaust and accelerates the comet. Why didn't we see the tail?
Perhaps it was very dusty and faint. Unfortunately, Oumuamua

is too far away for any follow-up observations. We will have to wait for the next interstellar visitor.

With a giant lunar telescope, we will be able to search for potential interstellar objects. With its immense light-gathering power, coupled with the lack of any lunar atmosphere, will enable us to study the formation of young solar systems. We will unveil their structure. At unprecedented clarity, we could track distant asteroids that are potential threats to us. The telescope resolving power will greatly exceed anything imaginable from the Earth, or indeed anything feasible from an independent space flyer. Ultimately it is size that counts. And the Moon is unique in providing us with a huge spacelike platform on which to build.

CHAPTER 9

Internationalization

I know not with what weapons World War III will be fought, but World War IV will be fought with sticks and stones.

—ALBERT EINSTEIN

Collaboration or Conflict

Let's try to place lunar exploration in an international context. We do not want to repeat the mistakes we have made on the Earth. The lunar saga is about to begin, with the first astronaut landings planned for 2024. There is still time to plan and to act.

The lunar village concept is futuristic, but the international space agencies are committed to going in this direction. The Moon offers a unique venue for many activities. One is astronomy. For the foreseeable future, the lunar platform is unparalleled on the Earth or even in space. Now is the time to persuade the space agencies that future lunar construction projects must include astronomy. Only then can we aspire to provide a grand vision for humanity with regard to one of the most important questions ever posed: What are our cosmic origins?

Practical details such as enforcement of low radio frequency bands exclusively for science need to be negotiated. We have achieved radio frequency clarity in certain bands on the Earth. We have made similar efforts to preserve the darkness of the night sky in certain areas. Why not protect the Moon? The lunar far side should be envisioned as a giant enclave for pursuing lunar science.

Reserving a permanently shadowed polar crater or two for astronomy is another worthwhile goal. We simply need to have some protective legislation. By providing a sufficiently detailed strategy, it is entirely feasible that the space agencies will be persuaded to make these options key elements in their forward planning rather than as highly compromised and descoped afterthoughts. Exploration of the Universe is one of the ultimate goals of humanity. Combining exploration with the more practical aspects of lunar development is a compelling vision for future decades.

The Worst-Case Scenario

Let's consider this futuristic scenario. It is 2184. There is an enormous Chinese mining colony at the lunar north pole. Pollution from the colony is driving friction with its lunar neighbors. There is a vast US-owned luxury hotel complex and city at the lunar south pole. It extends over hundreds of kilometers with an array of sports facilities, health spas, and high-end shops.

The lunar mining industry is booming. Exploration for rare minerals prevails in a wide range of terrain, from highlands to craters. It is the gold rush all over again. Wealthy Earth dwellers have homes and apartments in the major population centers. Lunar explorers, scientists, and workers generally feel excluded from the bustling downtown activity. They live in housing complexes on the outskirts. Huge entertainment complexes provide distraction. Both the Chinese and American colonies host booming spaceports and commercial centers. Russia has developed a giant lava tube discovered near the Marius Hills skylight. Here the first large terraformed lunar metropolis has been built. New Moscow is totally enclosed and protected by the lava tube. The outer surface is impervious to asteroid impacts. It can even resist nuclear explosions. Many inhabitants were born on the Moon, and few have even visited the Earth for vacation.

The lunar colonies have formed mutual trade associations that have developed close cooperation in security and economic infrastructures. Each colony is at a different stage of agitating for independence. Several are already demanding the right to self-government and control of taxation. New Moscow, by far the largest lunar city, with strong indigenous ties, has too much at stake to lose in any conflict. The megapolis is waiting to see how the situation evolves. Russia, by far the commercially weakest of the major terrestrial powers, is in no position to exert any dominant military or political role.

Of course, there is much potential investment and financial reward at stake. Neither the Chinese military nor the US banks and investors support the devolution of their lunar colonies. Stress inevitably mounts between the various stakeholders. The colonies have set up a mutual defense pact. Russia might opportunistically take sides as it hesitates between its terrestrial and lunar allies, inevitably with its own interests in mind. Unless cooler heads prevail, the stage is set for a dispute that could trigger Lunar War I.

It would be sad to see a rerun of terrestrial independence conflicts on the Moon. Without a strong legal framework, which has yet to be developed, territorial quarrels and trade wars will arise. A lunar legal constitution is urgently needed to handle colonial disputes and guide conflict resolution toward peaceful outcomes.

How do we cope with territorial claims on the Moon? Lunar exploration has inevitable military as well as commercial connections. Looking further ahead, we can see that, once significant commercial interests are at stake, we may have to cope with jurisdictional issues and military law enforcement. We can hope for an international lunar treaty, but so far this has been elusive.

Legal Treaties

The Outer Space Treaty (OST) was first signed in 1967 at the United Nations, and it had been ratified by 110 countries as of 2020.[1] Some 23 more countries were in the process of ratification. The treaty's main aim is to limit the launch of nuclear weapons into space. Use of the Moon is limited to peaceful purposes only. The Outer Space Treaty specifies that no nation may claim sovereignty of outer space, and its emphasis on access is clear: "All states shall have equally free access to Outer Space including the Moon and other celestial bodies. . . . Outer space is not subject to expropriation by claim of sovereignty, by means of use or occupation" (Article I, OST, 1967). According to the treaty, no country can lay claim to swathes of the Moon.

The principle is somewhat similar to the Antarctic Treaty of 1959, which established free scientific exploration and allowed research by any country on the Antarctic continent. It also prohibited commercial exploitation of Antarctic resources and any military activity on the continent.

The Outer Space Treaty does not ban military activities within space or the weaponization of space. One exception is the placement of weapons of mass destruction. The treaty has an environmental clause, namely, that exploration should avoid any harmful contamination. Even criminal law is encompassed by the treaty. Its enforcement, however, is an entirely different issue.

NASA has proposed an updating of the Outer Space Treaty. The Artemis Accords were signed in 2020 by the United States, the United Kingdom, Brazil, Canada, Australia, and a handful of minor partners. All of these nations are participating in NASA's Artemis program aimed at returning humans to the Moon by 2024 and further developing lunar exploration. There is no broad international agreement by which, for instance, operating licenses would be issued via the United Nations. The future of commercial exploitation, including mining operations and extraction of lunar resources, and how to handle possible conflicts remain to be decided.

Real Estate

Astronomers frequently collaborate with colleagues who work for other major space powers.[2] Joint international projects are commonplace. Currently astronomers share many facilities and many databases, and there is international access to the newest telescopes, both on the ground and in space.

The limited space on the Moon, especially prime space for developing infrastructure, is what should concern all prospective lunar scientists and engineers, including base developers, transport specialists, telescope builders, and especially lunar constructors and entrepreneurs. There are choice areas of lunar real estate. Many have simultaneous access to almost continuous sunlight in close proximity to perpetual darkness and cold, and mineral-rich mare. The combination is incredibly attractive, and not just for building telescopes. Unique manufacturing and futuristic life-prolonging medical operations can be carried out in such low-gravity environments.

There are other regions with choice mineral deposits where ancient asteroids crashed into the Moon. The potential resources are unimaginable, given the diverse lunar geology. Imagine all the billions of asteroids whose debris has accumulated over billions of years. Unlike most terrestrial asteroid hits, lunar asteroids did not burn up. Legal codes must be developed to enable commercial activities and establish the legal basis of mining claims and ownership of extracted ores.

Water may be the most valuable resource of all. Water is key to the rocket propellants needed for travel in the inner solar system, or even beyond. Crater resources of ice are mostly limited to the polar regions. There may be more widespread regolith reserves of water.

Rare element mining on the Moon will provide resources long after terrestrial mines are exhausted. But such extraction comes at a price. The mining industry is notorious for terrestrial pollution. How can this be avoided on the Moon? And who will be responsible for pollution-related costs?

Environmental issues must be addressed, as well as restrictions on the use of frequency bands for communication. Noninterference in mining operations needs to be clarified, as does enforcement of antipollution regulations, once they are established. The number of promising mining sites is equally limited. Who has priority on a particular site? First come, first served? Or will we succeed in setting up a framework in advance, as with the Antarctic treaty?

Space Forces

Plans for a new US Space Force as a sixth military service, on par with the Army, Navy, Air Force, Marines, and Coast Guard, were announced by US vice president Mike Pence in 2018. The motivation was to ensure America's dominance in space. "Previous administrations all but neglected the growing security threats emerging in space," Pence said. "Our adversaries have transformed space into a war-fighting domain already, and the United States will not shrink from this challenge." The Space Force is the first new US military agency to be established since the Second World War, and it has been endorsed by the Biden administration. The Space Force is specifically intended for the human return to the Moon. The aim is for the United States to combat potential threats from China and Russia. Presumably these countries will make similar plans. The stage is set for an armed presence on the Moon. Let's hope we succeed in demilitarizing the Moon by mutual agreement between the major players.

The Moon promises to become a crowded environment over the next decades. International coordination is still rudimentary and needs to be improved. Space is inevitably a military arena. Currently, sophisticated and expensive spy satellites monitor the skies from low Earth to geosynchronous orbits. The geosynchronous location is where the satellite remains at the same point in the sky. Its orbital time matches the Earth's rotation. It's the preferred place for spying.

Those satellites require protection, since antisatellite weapons such as interceptor spacecraft can destroy a satellite in space. The next step will surely be to monitor cislunar space, all the way to the Moon and its surrounding orbital space. The temptation will inevitably be to develop capabilities for satellite protection as well as for military intervention.

There are other dangers. The current Outer Space Treaty allows the use of military personnel for peaceful space activities, but such activities could be open to widely different interpretations. Protection will be needed against hostile forces, including space pirates and hostile spacecraft. We will be reliving the glorious period of continental discovery and exploration on the Earth, but in a totally new environment. Let's hope that we can learn from history.

Natural threats include meteoroid impacts of far greater danger on the Moon than on Earth, where all of the potentially dangerous space meteoritic grit burns up in our protective atmosphere. Rescue-and-recovery mission capability will be needed for the lunar installations. Meteoroids are not selective about where they strike. Advance warning systems and protection need to be an international effort.

Law enforcement will be the responsibility of the various national space forces. What is far from clear are the different roles of the various space agencies involved, including, notably, both the China National Space Administration (CNSA) and the National Aeronautics and Space Administration (NASA) in the United States. China will be establishing a lunar base over the same time frame as the United States. Like its US and Russian counterparts, NASA and Agency Roskosmos, the CNSA has a strong military heritage from the days of intercontinental ballistic missiles. The agencies have expressed strong interest in industrial development of the Moon. No doubt China and Russia will also be eager to establish their own space forces to police these activities in the not too distant future.

The expected growth of commercial activities will require legal enforcement to avoid seeding potential conflicts and chaos. There must be international legislation to adjudicate commercial disputes, adapted to the lunar context. Individual states' lack of sovereignty on lunar territory will leave the door open for commercial rivalries to expand. They must be reined in, if not actually controlled. One view is that only a military space force can legitimately operate on and around the Moon to defuse potential conflicts.

We could imagine that lunar mining eventually becomes critical for supplying terrestrial rare earths and power generation. If so, extraction of lunar resources would control the world economy. Mining disputes are inevitable, given the limited number of prime lunar sites. Some degree of internationally regulated law enforcement will be needed. We have yet to make the laws.

Looking further forward, the lunar space stations that are now being planned have an essential role to play in controlling human activities on the lunar surface. We want to avoid a replay of territorial disputes as well as anything like the terrestrial piracy wars. Historically, the breakdown of law and order has created immense problems for terrestrial society. Let's avoid this on the Moon.

Military Activities

Humanity must avoid war in space over lunar resources. Such a conflict would undoubtedly involve nuclear weapons and be as disastrous for the Earth as for the Moon. The Outer Space Treaty effectively demilitarizes the Moon. The agreement allows exploitation but not ownership. Unfortunately, the treaty has no enforcement provisions.

Disputes will inevitably arise. If one state reserves a site for prospecting and mining, how and for how long is this right maintained? How are demands for the use of adjacent sites to be allocated? Some sites are more desirable than others, such as regions with access to dark polar craters in permanent shadow, with ice deposits in the crater basin cold traps and available solar power from the illuminated polar crater rims. There surely will be an intense competition for the best sites, whether for mining or even for constructing telescopes.

We are used to competition on the Earth. Here the deepest pockets prevail, in accord with nationally or internationally agreed environmental constraints. And international law controls any potential disputes. We need to develop a framework for regulating lunar activity as well. Little has been done in the realm of outer space for more than two decades. Now, with the emerging new space race, it is especially urgent to set up an enforceable framework for regulatory control.

Criminal Law

Interesting legal questions will arise on the Moon. If an astronaut commits a crime on the Moon, what jurisdictions will be involved? We have similar issues on the Earth. A crime committed in an airplane flying over international waters, or on a cruise ship sailing through international waters, is regulated by international maritime law, which will consider several options. What was the jurisdiction at the last port of call, or the next port of call, or the final destination? Which jurisdiction prevails for lunar missions? Is the relevant jurisdiction determined by the registration of the spacecraft? This possibility opens up the prospect of legal scuffles if different national jurisdictions get involved. International maritime law provides a way of resolving these issues, but there is no lunar law yet.

On the Moon, the situation is inevitably more complex than the terrestrial circumstances covered by maritime law. There is no ownership of lunar territory, according to the Outer Space Treaty, which focuses on developing peaceful exploration of space. What happens when a serious crime is committed on the Moon? Should the jurisdiction of the nation that launched the spacecraft be considered responsible for the lunar voyager? If the crime is committed against a national from a different country, then there is no simple resolution. And if the crime is committed on a lunar base belonging to a third party, the situation becomes even murkier.

Spaceports

The Moon is envisaged as the launch point for crewed exploration of the solar system. Mars will no doubt be the prime destination for human travel a half-century from now. In the meantime, we will focus on the Moon. Rocket fuel is easily generated on the Moon, and liftoff of heavy loads is feasible under lunar gravity. A lunar spaceport would provide the ideal refueling and repair station for spacecraft designed to venture farther afield.

Launching spacecraft is a dangerous activity, and rocket exhaust often contaminates a large area. We would like to avoid this on the Moon as we prepare for interplanetary missions, or even just return to Earth. One solution is to launch from an orbiting lunar space station. This concept underlies NASA's Artemis program for initially sending shuttles from lunar orbit to the lunar surface. The focal point of Artemis will be a lunar orbiting space station and spaceport.

The long-term goal is the stars, but there are many challenges ahead if we are to meet that goal. One challenge will be equipping human crews to cope with the length of interstellar voyages. The physical and psychological challenges are extreme. The favored solution must be hibernation. Animals successfully hibernate for winters, and there are recorded examples of much longer hibernation periods. There is no biological reason why we should not be able to develop hibernation for humans. The stars beckon.

Lunar Transport

It is expensive to launch payloads to the Moon. The expense is driven by the cost of launching from the Earth, where the weight of fuel and hardware dwarfs that of the payload. NASA's space shuttle weighed in at 2,000 tons, but its effective payload was 30 tons. And it was launched just to near-Earth orbit.

Of course, factories could be established on the Moon to provide building materials and fuel supplies. But many tons of material will need to be transported from the Earth. It should be possible to cut the transport costs, as has certainly been achieved with terrestrial air transport. Can we develop cost-cutting strategies for space exploration?

Recyclable launchers are a partial solution. In fact, NASA's space shuttle greatly lowered the transport costs in the post-Apollo era. In current dollars, Apollo cost some $150 billion. The NASA space shuttle launches cost about $1 billion per flight. Launching to near-Earth orbit is relatively cheap: something like $10,000 per kilogram is the current cost to get to an altitude of 1,000 kilometers. A ticket on Richard Branson's Virgin Galactic space launcher will cost about $450,000 for a sub-orbital 90-minute space trip in 2022. Branson's passengers will get to an altitude of 50 miles, the notional beginning of space.

A midway point energetically would be geostationary orbit at 36,000 kilometers. Current charges for delivery by launch companies are about $50,000 per kilogram. This controls the cost of launching a typical communications satellite into a preferred orbit. From there, only a relatively small acceleration is needed to slip entirely out of the Earth's grasp. Arranging for a spaceport at geostationary orbit would be the ideal launch point and could be a way of further cutting the costs of lunar delivery. A rocket launched from Earth requires 100 times the weight in fuel per kilogram delivered to the Moon, but launching from space would dramatically ease the weight requirements of a spacecraft.

However, lunar spacecraft are expensive. Current estimates by private companies competing for NASA contracts allow for $1 million per delivered kilogram of payload to the lunar surface. This cost becomes excessive for delivery of a 10-ton payload, the current target for the first generation of lunar missions. The inflation-adjusted space transport costs have certainly come down over the last half a century, but they remain extremely high. Experts argue, however, that as ultraheavy recyclable spacecraft become available for lunar transport, the cost should fall to hundreds of dollars per delivered kilogram.

The first steps in commercial crewed space flight were taken in 2021. Elon Musk's SpaceX company pioneered commercial participation in crewed space flight and won a NASA contract for its Starship to land astronauts on the Moon in 2024. Two other aerospace companies competing for contracts for construction of space transporters had successful suborbital flights with civilian crews in 2021. Blue Origin founder Jeff Bezos flew in the New Shepard spacecraft with paying tourist passengers. A week earlier, Virgin Galactic founder Richard Branson had briefly blasted into suborbital space on his Unity spacecraft. Space tourism is underway.

Space Elevators

One exciting prospect is for a space elevator that could haul huge masses to heights of 50,000 miles above the Earth. From there, launch requirements to the Moon are dramatically eased. The idea of a space elevator was first concocted in 1895 by rocketry pioneer Konstantin Tsiolkovsky, who supposedly was inspired by the newly constructed Eiffel Tower. Modern discussions demonstrate its feasibility, at least in principle. The concept reads like science fiction, but the physics is sound.

A space elevator is built by starting with a space station in geostationary orbit. At a height of 20,000 miles, a satellite orbits at the same rate as the Earth spins. It is always visible from a fixed point on the ground. Many communications satellites are launched into such a geosynchronous orbit in order to enable continuous coverage of a location on the Earth. From the geostationary space station, a tether cable is dropped down to Earth. At the same time, a counterweight is deployed in the opposite direction. The tether to the counterweight is pulled up by the outward-directed centrifugal force. If the counterweight tether is long enough, any object deployed as the counterweight will be ejected out into space, above the escape velocity. The tether would need to be long; about another 20,000 miles would do nicely. The tether also needs to be constructed from light but strong material. Carbon fiber technology might provide a starting point for development.

So now we have a space elevator. Cargo carriers mount the tether to the space station, depositing cargo. Then the cargo is taken up the second cable to the launch point at the counterweight, 40,000 miles above sea level. From here, the centrifugal force propels a spacecraft containing the cargo containers at escape velocity. The cargo is en route for the journey to the Moon.

It does sound like science fiction, and perhaps it is. But such a space elevator does not violate the laws of physics. Nothing extraordinary is needed beyond some advancements in material science to ensure the tensile strength of the cable.

Building a space elevator for lunar launches is easier. The lower gravity of the Moon helps. A lunar space elevator could probably be built with a cable made from known advanced materials of adequate tensile strength. The tensile constraints are moderated in the case of the Moon. Making a long enough cable is another technical challenge. A cable of some 30,000 miles, longer than what is required for Earth launch, would be needed to get to the lunar orbiting docking port. But the result would be a low-cost system for mass delivery of lunar resources to Earth.

CHAPTER 10

The Next Century

The Earth is the cradle of humanity, but we cannot live forever
in a cradle. . . . Men are weak now, and yet they transform the
Earth's surface. In millions of years their might will increase to
the extent that they will change the surface of the Earth, its
oceans, the atmosphere, and themselves. They will control the
climate and the Solar System just as they control the Earth.
They will travel beyond the limits of our planetary system; they
will reach other Suns.

—KONSTANTIN TSIOLKOVSKY

It will be difficult enough to avoid disaster on planet Earth
in the next hundred years, let alone the next thousand, or
million. The human race shouldn't have all its eggs in one
basket, or on one planet. Let's hope we can avoid dropping
the basket until we have spread the load.

—STEPHEN HAWKING

A Glimpse of the Future

Driverless lunar rovers run along the highways that traverse the
lunar surface, carrying passengers who have completed their
work shifts. Large convoys leave the lunar cities and return
laden with minerals for processing in the factories. The highly

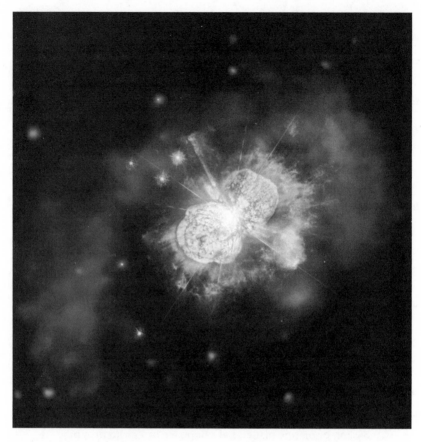

FIGURE 10. An unstable star destined to be a supernova in a century or two. Any exoplanets orbiting Eta Carinae, which is about 7,500 light-years away from Earth, are in imminent danger of destruction as this unstable star expands and eventually explodes. Our own Sun will suffer a similar but less violent fate about five billion years from now, expanding tens of thousands of years to burn the Earth to a crisp. The image of Eta Carinae is about 45 arcseconds across (about 1.6 light-years) and is a composite of separate exposures acquired by the WFC3/UVIS instrument on the Hubble Space Telescope (HST).

Image credit: The HST observations, taken in 2018, are in programs by N. Smith. https://www.nasa.gov/sites/default/files/thumbnails/image/archives_etacar.jpg.

automated factories assemble products destined for use on Earth. The preordered production line uses nanotechnology and 3-D printing. Humans supervise robot designers and guard against coding malfunctions. The manufacturing processes benefit from the low gravity and near-perfect vacuum environment. They deliver products whose synthesis is unfeasible on Earth.

The workers live in lunar cities based in giant lunar lava tubes. Unlimited energy is available, sourced by solar power and thermonuclear reactors. Gleaming skyscrapers abound. Residential taxes favor vertical developments, and these living spaces are coveted by city dwellers. The most magnificent sight in the sky is earthrise. It is available to all watchers of the sky. Terraforming large swathes of the Moon allows construction of international parks and leisure facilities. As the shifts end, the workers head for the large resort complexes built near the lunar poles, the regions with the most temperate climate.

Work is a short shuttle ride away from the residential developments. Food service is automated. Few need to cook. In the resorts, the guests have space for unlimited leisure and exercise. There are fine restaurants and theaters, coffee bars and nightclubs, exercise facilities and sports stadiums.

There is no lunar currency. All transactions are conducted through credits. Payments are issued for work shifts and utilized for leisure. There is a lunar prison with a handful of inmates. Pending appeal, most will be shipped back to Earth. With its mostly self-selected population, the Moon is a low-crime environment.

Births are encouraged. Hospitals and unlimited medical coverage are free for all residents. A new generation of residents born on the Moon has emerged. They thrive in the low-gravity, pollution-free surroundings. The athletic skills of Moon residents are superb, and lunar athletes regularly win contests with terrestrial colleagues. The clean environment keeps disease rates low, and life spans, already exceeding a century, are continually being extended by improved medical care and judicious use of transplants. Medical technology in the low-gravity environment vastly surpasses terrestrial possibilities for transplants of artificial organs. Immortality beckons. The University of the Moon hosts 10,000 students. Its primary focus is on interplanetary travel and robotics studies. Novel spacecraft are built as prototypes, and new types of rocket fuel are developed. Advanced robots are being designed to meet the challenges of travel to the nearest stars.

An accountable legal system has emerged. Crime is controlled by an international enforcement agency that maintains a peaceful coexistence among the denizens of the multicultural environment. Territorial disputes and mineral claims are resolved by adjudicating panels established and regulated by a lunar court of justice. Bonds underwritten by all countries that exploit the Moon secure investments and help to develop a major source of revenue from commercial activities. Pollution is strictly controlled. Insurance companies are deployed to cover the costs of any major accident on the Moon.

This scenario could describe the Moon one century from now, provided that we succeed in putting competition to one side and work to preserve the pristine nature of the lunar environment. If we fail, we can expect standoffs between nations making rival claims and disagreements on the adjudication process. There will be endless debates about jurisdiction. Ultimately, we might expect shows of force if no international treaty has been negotiated.

But rivalries and disputes can be managed. Antarctica is a local example of a vast shared territory, albeit the challenges on the frozen continent are far less demanding. Pollution there has largely been avoided. Global warming represents the ultimate threat to destroy Antarctica, but we may yet succeed in bringing this under control.

The Moon presents other challenges. Global warming is irrelevant because there is no lunar atmosphere to drive the greenhouse effect that feeds on terrestrial carbon production. Fossil fuel burning and related human activities will not pose a lunar threat. The temperatures on the lunar surface already range between unimaginable extremes, but we know how to cope with them. Our biggest lunar challenges, besides the hazards of daily life on the Moon, will be man-made pollution and the inevitable tension between commercial exploitation and scientific use. The greatest threat may be embedded in the rivalry for production of the vital metal ores needed to feed rare earth elements and semiconductors to terrestrial and lunar industries.

Of course, we might have imagined similar issues prevailing at the onset of the industrial revolution in the nineteenth century. Air, soil, and water pollution increased, and wild habitats were dramatically reduced. And poets were inspired. Most notably, William Blake wondered:

And did those feet in ancient time
Walk upon England's mountains green . . .
And was Jerusalem builded here
Among these dark Satanic Mills?

But the voice of reform was muted. Thanks to the huge increase in the efficiency of producing goods and providing employment, the economic benefits were soon overwhelmingly positive. The social issues would mature more slowly. Only very gradually did living and welfare conditions improve. But eventually the industrial transition, at least in developed countries, resulted in a mostly harmonious world order.

There will be an interesting debate on self-regulation as lunar exploration progresses into the construction of lunar habitats. Hopefully, we will have learned enough from the mistakes that are being so painstakingly corrected on the Earth. Commercial lunar activities must be controlled. Mining activities can be shielded, and residential construction and factories can be isolated. Doing so will no doubt be expensive, but the immense potential rewards will pave the way toward an accommodation by international consensus that preserves the pristine state of most of the Moon. It is currently virgin territory, so we should have time to act.

Lunar Vision

Current initiatives are more technical and commercial than scientific. Such goals certainly set the scene. Yet humanity ought to adopt a broader vision. Exploration of our frontiers has defined human progress over the millennia. The final frontiers are our future. They hold the secret to whatever mysteries there are to be unveiled in the Universe. It is an undeniably rich menu, one that has excited the human imagination over the millennia. The Moon can take us there.

We need to look upward and outward. Unless we start planning now, the lunar adventure will lack an exceptional asset—a lunar observatory to host a family of giant telescopes spanning radio, optical, and infrared wavelengths. Such telescopes would be uniquely poised to answer one of humanity's most profound questions: What are our cosmic origins?

To be sure, observations from Earth and from orbiting satellites are impressive. The Sloan Digital Sky Survey mapped more than a million galaxies. Larger surveys have mapped 100 million galaxies. Underway in the next five years are new surveys that promise to deliver data on several billion galaxies. These galaxies formed millions of years after inflation occurred, and their distribution in space holds secrets about the beginning. But the ultimate question relies on how to study the elusive gas clouds that preceded these galaxies. An understanding of what happened even before these clouds formed is the missing element of the cosmic jigsaw puzzle. We need to penetrate the dark ages.

We explore the past by studying fossils. Whether relics of Stonehenge or the Roman empire, fossils are our guides to long-faded civilizations. Even before civilizations, there were clues. Scientists explore the tracers of evolution, the origins of cellular life forms, the clues to survival and reproduction. The genetic code contains hints of the origins of life and its spread around the Earth. Radioactive dating probes our past on timescales of tens of thousands of years up to billions of years. We study our surroundings and search for the most primitive indications of our origins.

For the Universe, the first clouds are our most primitive fossils. If only we can read their message, built-in memories will illuminate the past. Not all clouds were created equal. Our galaxy formed long ago from millions of such clouds. There are tens of billions of galaxies that we detect with our largest telescopes. All were built from assemblies of small clouds. Looking far back in time allows us to map these clouds, if our observing techniques can meet the challenge. With trillions of clouds to sort through, there will be relic signatures that we struggle to decipher. These will tell us about subjects as diverse in scale as the nature of gravity and the role of the quantum theory in determining the seeds of structure formation.

There is a way forward. The new frontier in cosmology lies long ago, before the first stars formed. Current proposals for lunar development must include exploitation of our best chance to glimpse the beginnings of the Universe: via signals that have traveled the farthest in our expanding Universe. The signals represent the dark ages—the first few hundred million years after the Big Bang, before the first stars formed. We know almost nothing about our dark past. It is territory that is ripe for harvest.

To gain the needed precision to tackle this challenge, we must look beyond the billions of observable galaxies. We must look to their building blocks: the trillions of dark clouds of hydrogen gas. Here is the site of our cosmic origins. The 21-centimeter radio signal from atomic hydrogen allows us to map these extremely remote hydrogen clouds. The distant signals are stretched by the expansion of space from a wavelength of 21 centimeters to around 10 meters. Detection represents a real challenge.

Such subtle distortions of radio waves from dark-ages hydrogen clouds cannot be detected by current instruments on Earth. At 10-meter wavelengths—that is, a frequency of only 30 megahertz—the Earth's ionosphere renders signals unacceptably noisy. Earth-bound radio telescopes also encounter too much interference from electromagnetic pollution caused by human activity, such as maritime communication and shortwave broadcasting. The dark ages are virtually inaccessible from the terrestrial perspective. The radio sky is too bright.

We need to go beyond the Earth. The far side of the Moon is the best place in the inner solar system from which to monitor low-frequency radio waves. These are the only way of detecting the faint "fingerprints" that the Big Bang left on the cosmos. Only from the far side of the Moon—with no ionosphere and shielded from Earth-related interference—could we hope to spot these dim shadows.

Cosmic inflation at the beginning of the Universe imprinted a tiny distortion on the clouds' distribution, in shadow against the cosmic microwave background. This provides a unique means to verify or falsify theories of inflation. It is the only certain signal from the beginning of time. Have scientists settled on too simple a model of the early stages of the Universe? We need to study these feeble signals to learn how inflation proceeded in the first trillionth of a trillionth of a trillionth of a second after the beginning.

A radio array able to capture these data would use millions of simple radio antennae deployed over an area that is 100 kilometers or more across on the Moon's far side. These would be deployed by lunar rovers, coherently connected by laser beams and operated by a combination of humans and robots. Orbiting lunar satellites would relay signals back to Earth. The view from there of the beginning of the Universe would be as good as it ever could be.

But there is far more to the search than the radio wave domain. Radio telescopes represent just the initial phase of lunar astronomy. We can do so much more. Infrared telescopes of unprecedented scale could be built in the permanent shadow of the cold craters near the lunar south pole. Here temperatures as low as 240 degrees Centigrade below zero, or 30 degrees Kelvin, have been measured. That is ideal for building infrared detectors. There is no atmosphere to absorb radiation and block signals. These telescopes will provide the only pathway to spectroscopically probe the atmospheres of millions of nearby rocky exoplanets. We will be able to search for robust signatures of alien life. Giant lunar telescopes could yield the first resolved images of nearby exoplanets.

Atmospheric signatures from distant planets might carry the dimmest and most cursory glint of an alien Earth-like planet vastly more advanced than our Earth. We are convinced that this is an option, although we have not the slightest idea of a possible detection rate. We simply must look. Our confidence that this idea is a sensible basis for searches comes from the simple fact that our Sun is just a middle-aged star. Half of the planetary systems have a billion-year head start on the Earth, or even more. Of course, we cannot begin to compute their survival odds. But fate would be extremely capricious were we alone in this vast universe. So we have to search.

The exact signal that we are seeking is a matter of continuing speculation. Advanced civilizations require vast amounts of energy, some of which may leak out from remote exoplanets. More sadly, there may be radioactive debris from catastrophic wars. There may be beacons, such as rapidly spinning neutron stars that are beaming coded radio signals to remote observers.

The most advanced species would most likely master self-propagating robotic spacecraft. These voyagers of the future would easily penetrate the awesome distances and survive the challenging travel times through interstellar space. Loaded with our genetic code, the spacecraft would propagate our species throughout the galaxy. Cryogenic advances might even lead to quasi-immortality during such long voyages. The first species to master such technology would have a head start. There would be propagation of the persistence of memory and culture. The entire Universe of galaxies might be abuzz with explorers and traders. This is a vision, one of many possible outcomes. We have not the slightest idea of what to expect in reality. We have to look.

Searches could be serendipitous or targeted; either type would have benefits for cosmology. Giant telescopes would also capture the images of the first stars in the Universe. We cannot look for planets at such huge distances, but we would be able to study how and when the chemical elements formed. They are the most fundamental precursors for life. A lunar platform will provide the basis for an observatory capable of rising to all of these challenges. But are such searches realistic? The cost would be immense.

A look at the history of space telescopes gives us reason to be optimistic. These are very expensive operations. The Hubble Space Telescope would most likely never have been launched were it not for the NASA budget that prioritized the space shuttle and the International Space Station. Launchers were made available. Throwing in the launcher system and the International Space Station for which the space shuttle was designed, I estimate that all space telescopes, Hubble included, cost no more than some 5 percent of the total space budget. We have been there and done that. We could do it again in response to the new lunar vision.

This gives us prospective hope for a lunar observatory. It will be expensive, but the lunar infrastructure will be far more costly. An observatory would cost a small percentage of the cost of the vast lunar transport system and associated developments that are being planned. An estimate of infrastructure costs would include payload delivery, construction, and deployment. Compared to these costs, telescopes are cheap, and the science returns from observatories on lunar sites are potentially huge. We should be inspired by the glory days of space travel now that the vast horizons of space exploration are beckoning again. The Moon will be our base for contact with the remotest depths of the Universe. It will be the ultimate staging point in our search for traces of extraterrestrial life.

We will launch new ventures to plan habitats in space and expand our footprint beyond the Earth. It will all be so much more invigorating than Earth-based space exploration, and profoundly stimulating. We will simultaneously look to the skies and connect with the deepest of cosmic questions. In previous centuries, our cosmic visions were led by explorers and poets. Today it is the turn of scientists to jump into the fray, to inspire and initiate the next stages in the progress of humanity.

Lunar hyperscopes will not be built for at least several decades, but I believe that they are inevitable. Imaging planets in distant solar systems is an irresistible goal. Our desire for exploration is too strong to resist. One challenge is that the telescope designers of today will have long retired before such megaprojects are completed. A long-term strategy is essential. Let's not forget that the architects of the greatest medieval cathedrals did not live to see their visions achieve reality. They too had compelling arguments for at least beginning the work. We need to make a similarly convincing case to our political masters.

What Lies Ahead

Current science planning is short-term, looking only decades ahead, at best. We should not neglect the unique opportunities that will be offered by lunar telescopes. Astronomers and the space agencies should develop and promote such adventurous ideas now, while lunar plans are still in their infancy. Producing rocket fuel from the lunar ice deposits and profiting from space tourists and mining operations are grand aspirations. But if we really want to challenge the limits of human exploration, we should seek out the beginnings of the Universe. And any neighbors. Only the Moon gives us that opportunity over the next century.[1]

In the immediate future, NASA's Artemis mission will establish a lunar base camp, beginning with the next series of astronaut landings in 2024. A high priority will be science, which will be focused on understanding the origin of the Moon and using the lunar surface to study its asteroid impact history and the record of the ancient Sun. It will be like reading an open book, thanks to the lack of atmospheric weathering, erosion, and any plate tectonics or major surface reshaping. A major goal will also be to observe the Universe from a unique location. Novel and powerful telescopes are planned. Far-side, low-frequency radio telescopes are likely be the first to be developed. Infrared telescopes are destined to take advantage of the unique observing conditions on the atmosphere-free Moon. Other space agencies will not be far behind.

These lunar developments promise to lead a major venture into the unknown. Our charting of new frontiers in space will best succeed if piloted by searches with the largest telescopes. Lunar observatories should be a vital element of future planning. Exquisite resolution from an atmosphere-free platform can be achieved only by hyperscopes on the Moon. These will help us chart the future of cosmic exploration. Exoplanets will be unveiled. Our deepest questions about life in the cosmos will be seriously addressed.

There are powerful commercial drivers for lunar exploration, centered on rare earths mining and eventually on tourism. Lunar ice will certainly be an immense reservoir for rocket fuel for travel to Earth and beyond. As the infrastructure for these industries is developed, it seems reasonable to anticipate that a compelling case can be made for incorporating science as we plan ahead. We need to plan for the future. The search for alien life is an inspiring goal. Understanding our beginnings in the very early Universe is science's greatest challenge.

Terrestrial fuel supplies of oil, liquid gas, and even coal will be exhausted within centuries, according to our best current estimates. Clean alternatives are needed. Tapping the very source of energy that drives the Sun is the ultimate power challenge. The Moon is a prospective station for collecting solar power. Microwave links could beam the power to wherever it is needed, whether back to Earth or into space.

There are more futuristic energy resources. The Moon might play an indispensable role in terrestrial energy needs, as it provides large amounts of helium-3. Use of this isotope of helium for thermonuclear energy is a very long-term vision. The technology is far from proven, and fulfillment is certainly centuries away. But it may well be in our future. The fusion of hydrogen and helium-3 could provide a clean source of energy, with no radioactive contamination. Exploration and exploitation of the Moon will facilitate such solutions to future energy needs.

The Moon beckons. Humanity's next major step forward will be to Earth's nearest neighbor. There simply is no alternative. The risks are too high, given current technology, for crewed travel to Mars, which is impractical for the foreseeable future. The journey would simply be too hazardous for human exploration or eventual installation of habitations. The physiological and neurological risks are extreme, and astronauts would arrive after a six- to nine-month journey barely capable of working. We will eventually figure out how to overcome these challenges. It is partly a question of shielding, partly a question of designing self-repairing biological systems, and no doubt partly a matter of developing new solutions that we can only dream about. Lunar bases will enable space travel throughout the solar system.

In the meantime, Mars is not being neglected. It will be abuzz with activities as several countries launch orbiters and rovers from 2021 onward. Mars is certainly the first major goal once we leave the Earth-Moon system. Robotic exploration of Mars is pressing ahead with renewed vigor. Mars is where fossils of long-extinct life may be buried from the intriguing and unexplored past of a few billion years ago. At that time Mars had a terrestrial-type atmosphere. We see evidence of water-driven erosion, dried-out river tributaries and deltas, and extensive residual ice caps on Mars. However, human exploration of Mars is far in the future.

The Moon is the next step for crewed ventures in space. But the sequel to establishing a base on the Moon will inevitably be: *Next stop, Mars!* Martian exploration will come after we are securely established on the Moon.

There are inevitable consequences of such plans for the lunar program. The space agencies of the leading world powers currently have a unified vision of lunar exploration and base building that is largely driven by commercial interests, not to mention by less explicitly stated military ambitions. Perhaps an equally important concern is control of pollution. But there are parallel scientific directions that we must not allow to be completely left aside as an afterthought. The unexplored frontiers of the Universe will be probed. This means opening up the dark ages and even earlier, going back to the first 100,000 years of the Universe. Now is the time to act as we enter into the grand vision of the future that is best summarized as *Back to the Moon!*

We have not set the best of examples close to home. We should not randomly pollute the lunar environment as we have done so devastatingly on Earth, with little thought of the consequences for future generations. The oceans are polluted, the atmosphere is polluted, and the extraction of mineral resources pollutes the environment. The global climate is warming, and the future century on Earth looks bleak.

The good news is that there is time to control these problems if we act soon. Many are aware of this challenge and are actively campaigning for a new strategy. There is cause for optimism if international rivalries can be set aside. But now is also the beginning of lunar development.

Above all, let's not re-create the mistakes we have made on Earth. Development and colonization of the Moon requires a long journey that will take perhaps half a century to develop, and centuries to complete. However, our rate of technological progress is staggering. Now is the time to set up the framework that will preserve the Moon as a safe environment for our grandchildren. Now is the time to initiate lunar-based science programs that will unfold the secrets of the beginning of the Universe. We will be able to evaluate the place of humanity in this vast and fragile cosmos of ours. Such ambitions will play a crucial role in guiding space exploration for the foreseeable future.

We will do science on the Moon, in addition to commercial activities. The latter provides the resources and the infrastructure, but it is scientific exploration that will take humanity to a higher level. We can unlock the secrets of the Universe thanks to the telescopes that we will construct on the Moon.

NOTES

Chapter 1. The New Space Race

1. Pre-Apollo, there were ten uncrewed NASA probes between 1964 and 1968. The successful launches included three Ranger hard impacts between 1964 and 1965 and five Surveyor soft landings between 1966 and 1967. The first robotic spacecraft to land on the Moon was Russia's Lunokhod 1 in 1970. Remotely controlled from the Earth, Lunokhod 1, along with its successor missions, was designed to image the lunar surface and study lunar regolith. Future Russian lunar missions were robotic.

The launcher for the Apollo program, Saturn V, carried some 140 tons to low-Earth orbit, with a third stage that transported the astronauts (and propellant) to the Moon. Saturn V was 300 tons when fully fueled, with about 10 percent of its liftoff weight deliverable to the Moon. After Apollo 17—the last crewed lunar lander to set down, in 1972—funding for the space program slowly, but dramatically, decreased. In the absence of political enthusiasm, funding was directed at other priorities, and there were no more crewed ventures to the Moon. Over recent decades, NASA's historical funding level has stayed at about half of 1 percent of the US federal budget. At the Apollo peak, its share of the federal budget was nearly ten times larger.

2. The Moon has not been entirely neglected since the glory days of Apollo. Several countries have been involved in making some thirty robotic landings—most of them soft impacts on the near side of the Moon—that have prepared the way for future crewed missions. A range of scientific experiments are being carried out in China's lunar program, Chang'e 4. Its rover Yutu-2 is exploring the lunar soil and has planted seeds whose growth will be studied. The lander craft beams data back to a small relay satellite orbiting the Moon. India has also been developing plans for lunar surface exploration. The first Indian lunar orbiter mission, Chandrayaan-2, was launched from the Sriharikota space station in 2019. Equipped with a lander and a rover, the launch vehicle was India's most powerful rocket, a three-stage monster weighing some 640 tons. Its lunar rover, called Pragyan (Sanskrit for "wisdom"), had a range of half a kilometer. The lander most likely crashed. However, terrain mapping for ice deposits and minerals is an ongoing objective of the orbiter. India was planning a

sequel, with a robotic lunar orbiter, lander, and rover to land near the lunar south pole in 2022.

3. In 2021, NASA selected SpaceX, headed by Elon Musk, for lunar delivery payload development. The competing companies were Blue Origin, owned by Jeff Bezos, and Dynetics, a smaller company that is part of the Leidos technology company. Launches of robotic, commercially built landers are expected to begin as soon as 2022, including a lunar resource prospector. These landers will lead the way to human exploration of the Moon. The post-Apollo era of space exploration will begin in earnest with the launch of the Orion crew capsule, which is being built by the European Space Agency. The first uncrewed test flight is scheduled in a joint ESA/NASA mission in 2023. The ESA is providing the propulsion module to give the final nudge into lunar orbit.

The new NASA generation of lunar landers is named after Artemis, Apollo's twin sister and goddess of the Moon. The Artemis program of lunar exploration by crewed lunar launches and landers is scheduled to begin in the mid-2020s. Crewed shuttles will land on the lunar surface from the Lunar Gateway, the orbiting space station that will suffice to take us back to the Moon in a meaningful way. The orbital platform, which will be capable of launching spacecraft to and around the Moon, will incorporate a crewed lunar-orbiting space station designed to host a fleet of crewed spacecraft. International partners have been lined up, including the European Space Agency, Russia's Agency Roskosmos, Japan's space agency (JAXA), and the Canadian Space Agency (CASA). Occurring in parallel will be launches to the Moon by CNSA, the Chinese national space administration.

4. There are three components of modern spacecraft that facilitate lunar exploration: an orbiter to survey the landscape, a rover to take soil samples, and the primary spacecraft lander. A series of CNSA lunar exploration launches are planned over the next decade. Chang'e 6 in 2024 will investigate the topography, composition, and subsurface structure of the landing site, and it will return south polar samples to Earth. Chang'e 7 will explore mineral resources at the lunar south pole in 2025 with an orbiter, a lander, and a rover. In 2027, Chang'e 8 will verify the utilization and development of resources with a lander and a rover. A flying detector probe will transport a small sealed ecosystem.

To explore the regolith around the lunar south pole the Russian space agency ROSCOSMOS will send a robotic mission, Luna-25, to the Moon in 2022. Successor Luna missions will map the lunar surface and examine the lunar regolith. In the mid-2020s, Luna-27 will drill into the lunar crust searching for signs of water in an effort to find out whether there is enough to support a future crewed outpost. Russia is building ultraheavy-load spacecraft for future crewed lunar missions. The Don launcher will be able to carry a 130-ton payload to low-Earth orbit and a deliverable payload of up to 32 tons to the Moon by 2029. This is competitive with the payload of NASA's planned ultraheavy-load launcher and with China's Long March 9.

In the next decade Russia will send its first cosmonauts on missions to orbit the Moon by 2029. Sometime in the early 2030s it is anticipated that Russians will for the first time set foot on the Moon when cosmonauts touch down on the Moon to construct a base near the lunar south pole. Japan's first lunar lander and rover is expected to be launched in the 2020s. Subsequent launches will also include a lunar sample return mission, along with an advanced lander for future human missions to the Moon. The Japanese Space Exploration Agency will join the ESA and NASA in developing a lunar base for scientific research and environmental studies. India is also developing its own crewed spacecraft. The Indian Space Research Organization hopes to launch such a space mission to the Moon in the 2020s.

5. The dangers of radiation exposure vary depending on the age and gender of the exposed person. The typical dosage is low on the Earth because we are protected by the atmosphere. The average dosage per person is only about one-third of a rem per year. That corresponds to the biological effect of the deposit of 100 ergs of X-rays in a kilogram of human tissue. A chest X-ray delivers typically about one-thirtieth of this dosage.

One of the principal calibrations in risk estimation has been the survival rate of Japanese atom bomb survivors. The maximum limit by international standards is about 0.1 sieverts in the usual international limits of exposure to X or gamma rays. The maximum radiation limit is 10 rems, where the rem represents the equivalent biological effect of the deposit of 100 ergs of gamma ray energy in a kilogram of human tissue. That's 0.1 of a sievert. One sievert measures the amount of radioactive radiation absorbed by a human body and is equivalent to one joule of energy per kilogram of mass. Even one-tenth of a sievert is believed to double the cancer risk over the next twenty years after exposure. When I was a child, shoe stores routinely used X-ray machines to measure customers' feet for shoes, delivering exposures of about one-tenth of this limit in twenty seconds.

Higher risk levels are expected of astronauts, as with test pilots. NASA's limit for low-Earth orbit trips to ISS is currently about five times higher for astronauts than it is for other people. A recent report suggested that the lifetime maximum permissible should be 600 rems, as calculated for the highest-risk healthy astronaut, considered to be a thirty-five-year-old woman for the purpose of the estimate. Cosmonauts on the long trip to Mars would greatly exceed this exposure, and such missions are unlikely to be undertaken until we develop improved radiation shielding.

6. The Japanese missions were the first spacecraft to return asteroid samples to Earth, in 2010, when spacecraft Hayabusa 1 (Japanese for "falcon") landed on asteroid 25143 Itokawa in 2005 to collect surface material. The sample mass of microscopic grains was minimal, about a milligram. The Japanese space agency JAXA followed

up with the Hayabusa 2 spacecraft, which was launched in 2014 and landed on asteroid Ryugu in 2019, 300 million miles away, to collect rock samples. With five grams of asteroid samples collected, Hayabusa 2 returned to Earth in late 2020, landing near Woomera in the Australian desert.

7. The rare earth elements that we expect to be mined on the Moon would include neodymium, cerium, dysprosium, europium, promethium, holmium, praseodymium, erbium, terbium, yttrium, niobium, samarium, scandium, gadolinium, lanthanum, and tantalum, as well as titanium and zirconium. Many of these have unique industrial applications, including telecommunications and computing.

8. Helium-3 consists of two protons and one neutron, whereas the dominant stable form of helium is of atomic mass four and consists of two protons and two neutrons. The international ITER (International Thermonuclear Experimental Reactor) project is pursuing thermonuclear fusion of deuterium and tritium for developing the first fusion reactors. The goal is to build the first sustainable controlled thermonuclear fusion reactors by 2050. The technology is known, and there should be no roadblocks to achieving this goal. The use of helium-3 allows for a much cleaner fusion environment, as far fewer neutrons are produced as by-products.

Helium-3 is known for permitting extreme cooling, which allows superfluids to be produced. The superfluid is a transition of a fluid to a state with no viscosity or friction. Imagine a spinning fly wheel. Air friction slows it down. In a superfluid, a spinning wheel spins virtually forever. A sort of perpetual motion machine might be constructed, one of science's oldest dreams, though it is unlikely to be of any practical use for energy extraction in the foreseeable future.

9. A futuristic lunar habitation has been proposed by the European Space Agency as a means of providing a focus for developing robotic and astronaut activities by all interested nations. In its 2020 engineering study "ESA Engineers Assess Moon Village Habitat," the ESA proposed a four-story building with shielding by locally engineered regolith-based bricks. Other countries, including China, the United States, Russia, and India, are considering similar activities. Both NASA and the ESA have extensive simulation projects that train astronauts for living on and exploring the lunar surface.

International space agencies are assessing various aspects of lunar infrastructure, including human habitats, laboratories, support for extra-vehicular activities, pressurized vehicles, and surface transport. The shopping list is long. We will need to construct power systems, life support systems, factories, and bioreactors, using regolith materials. Other measures might include transportation of prototypes and laboratory-tested elements along with simulations of realistic environments and mass rescue procedures.

To facilitate habitability we will need to develop emergency services, find ways to meet medical needs, and design escape shelters and inflatable building extensions. Transport and deployment of equipment from Earth will require enhanced ultraheavy payloads. Delivery capacity is expected to be managed by the private sector through a commercial Amazon-like delivery system. Key, however, will be production of construction elements in local factories and, for the scientists, construction of large observatories.

10. Permanently shadowed polar craters are high-priority sites for a range of activities. High rims provide an almost unlimited source of perpetual solar power. We will need to make use of in-situ manufacturing and 3-D printing for elements and structures. Human on-site activity will permit human and robotic tasks to be powerfully combined—an essential accomplishment given the time delay for communication to Earth. We will need to address the challenges of communications and computing. Data storage capacity is central. Preparation will take many years, but enormous progress will be made in the next decade before we begin serious construction. Eventually we will make the great leap to the Moon. We will need to perfect a phase of telerobotic precursor demonstrations. Engineers will simulate telescope deployment and operations using analog digital devices.

Optimal sites for optical and infrared telescopes are in the shadowed basins of polar craters, many of which are in permanent shadow. At an ambient temperature as low as 30 degrees Kelvin, these basins contain ice and are prime sites for mining water as well as for locating telescope sites. One could build 100-meter-class telescopes in such locations, far exceeding the apertures of any terrestrial rivals.

More grandiose schemes include a hyperscope that envisages using a naturally dark polar crater with a diameter of several kilometers. Its surface would be covered with a network of telescope mirrors all pointing to a camera located at the central focus point and strung on wires spanning the crater rim. Novel quantum techniques will need to be developed to combine the telescope beams.

11. Analogous features to the giant lunar lava tubes are found on Earth, but on a much smaller scale—for example, in the Craters of the Moon National Monument in Idaho. A large lava delta produced the Kilauea volcano in Hawaii and sank into the sea in 2016, exposing the mouth of a large lava tube. An analysis of large lava tubes on the Moon reveals that they are a thousand times wider than those typically found on Earth. Some of the lunar tubes are large enough that they could contain and shield entire cities.

12. Looking back into the distant Universe, a million years after the beginning, one literally is in the dark. There are no galaxies and no stars. There is only hydrogen gas, in the form of clouds. The clouds are colder than the primordial fireball fossil radiation, seen as the cosmic microwave background. We call this the dark ages.

As the Universe expanded, the density dropped. Initially, after the last scatterings of matter, the radiation began to thermally separate from the hydrogen. Once the scatterings were over, atoms cooled more rapidly than photons, and when they became colder than the background radiation, they left a shadow. We can look for that shadow.

Some effects of the terrestrial ionosphere are beneficial: radio wave transmission at long wavelengths on the Earth is facilitated, since waves bounce off the ionosphere. For radio astronomy, however, the effects are disastrous, as low-frequency signals from space are scattered and deflected. Low-frequency radio astronomy telescopes will be located on the lunar far side, where terrestrial radio interference is minimal.

One example of what we could do in a really radio-quiet environment would involve hydrogen atoms. The predominant constituent of the interstellar medium in our galaxy, hydrogen atoms emit or absorb at a radio wavelength of 21 centimeters. Hydrogen clouds are the precursors of galaxies, and they dominated the early Universe before the epoch of galaxy formation. To observe the Universe before the first galaxies formed, when the Universe was completely dark and fifty times smaller than its current size, we need to study hydrogen clouds. The hydrogen wavelength shifts and lengthens correspondingly as we look back in time, to a wavelength of 10 meters for ancient hydrogen.

Chapter 2. Digging Deep on the Moon

1. Planetesimals fused into larger and larger rocks. The ice and tiny dust particles acted together as a sort of glue to speed along the adhesion process. On the smallest scales, the debris around the forming planets formed meteorites, asteroids, and moons. Rocks predominated. One of the most striking properties of the solar system is the difference between the inner rocky planets, Mercury, Earth, Venus, and Mars, and the outer icy giants. The hydrogen-rich ices dominate over the silicate-like rocks. The net effect is that the volatile material in the outer solar system accumulated to form the giant gaseous planets. The most prominent planet in our solar system is Jupiter, a gas giant of some 300 Earth masses with a rocky core that contains about 5 percent of its total mass, or about 15 Earth masses.

2. Here is why we are sure that the Moon formed in close proximity to the Earth. A fundamental law of physics is that the angular momentum of a system must be conserved. The Earth loses spin because of the tides, and the lunar orbit takes up the difference in angular momentum. The Moon moved outward, away from the Earth. We observe that there is so much angular momentum in the Moon's present orbit that it must have formed close in. Initially it was just some two Earth radii away, where it formed billions of years ago. The Moon ended up in its present location of about 60 Earth radii, or nearly one-quarter of a million miles, from its parent planet.

3. Some 10 degrees Kelvin is achieved in dense interstellar molecular clouds, relying as they do on the low-lying electron energies of carbon and other heavier species.

Such energy levels allow more effective gas cooling to lower temperatures. These atoms have electrons in low-lying energy levels. When excited by collisions with hydrogen atoms, these electrons jump up a level, then fall back down again by radiating infrared photons. This radiation freely leaves the cloud, and the cloud cools down.

In the first clouds, there was no carbon. A sprinkling of hydrogen molecules formed. The hydrogen molecule's lowest energy level came from its spin. According to the quantum theory, spin comes in discrete amounts or energy levels. The lowest molecular energy corresponds to 512 degrees Kelvin. This is much higher than the typical atomic energy levels for carbon, or even silicon. Cooling still occurs, but it is now much less effective. These molecules were needed to cool the first clouds down sufficiently to fragment into stars. The traces of molecular hydrogen acted as a special coolant that was really useful in the early universe, which was uncontaminated. Clouds consist of atomic hydrogen, but there are a few leftover electrons from earlier hotter periods.

A hydrogen atom catches a free electron to form a molecule that we refer to as hydrogen-minus. This in turn undergoes an exchange reaction with another hydrogen atom to form a hydrogen molecule. This process returns the borrowed electron. In this way about one molecule of hydrogen is created for every 1,000 hydrogen atoms. This is the only coolant, but it's enough, even if not all first clouds form stars. If the clouds are too small, they will be too cold to cool any further and contract under the pull of gravity. They cannot radiate away the extra heat acquired from gravitational contraction. The first clouds have masses of at least one million times the mass of the Sun. These are warm enough to excite molecules and cool down. And they will aggregate to form the first galaxies. Many are much smaller than typical galaxies today.

4. The first stars are astronomers' fossils. We have not seen them directly, as they were so short-lived. Their surroundings were littered with debris—the heavy elements ejected by exploding stars, such as carbon, oxygen, and iron. This debris ended up being recycled into a new population of clouds that in turn formed the next generation of stars. These long-lived second-generation stars are still around today. We see them scattered throughout the halos of galaxies and even in nearby dwarf galaxies. Their heavy-element content is the fossilized relic of the first stars.

5. The fluctuations in matter were still able to develop as the expansion proceeded. At later times the radiation was subdominant. Still, the radiation remained tightly coupled to matter by the scattering off of the electrons. It decoupled some 380,000 years after the beginning when the radiation temperature dropped below 33,000 degrees Kelvin. This meant that too few photons remained that were capable of ionizing hydrogen. The atomic era began.

How deeply can we dig into the Universe? We will look back to long before there were any galaxies. Galaxies are key elements of the Universe. They probe cosmology and control our existence. The broad lines of their formation and evolution are clear. From infinitesimal density fluctuations present very near the beginning of the Universe smaller structures developed to eventually form galaxies over billions of years. It's bottom-up—that's our theory.

The process of fluctuation growth was slow for two reasons. First, for the first 10,000 years the Universe was so hot that the dominant radiation field inhibited matter from condensing. The repulsive pressure of the radiation was much larger than the attractive force of gravity. Second, the Universe is expanding. Consequently, all the matter in the Universe, which initially is almost perfectly uniform, is also expanding. The expansion greatly weakens the pull of gravity. It's like running on an ever-expanding racetrack. The growth of fluctuations in the matter density is reduced. When the Universe cooled sufficiently 380,000 years after the beginning so that the matter was almost entirely atomic, there were very few electrons around to transmit the tendency of the pressure of the radiation to counter gravity. The brakes on clouds were removed. Once the scattering ceased, cold gas clouds could begin to condense as tiny traces of hydrogen molecules boosted the gas cooling.

6. According to the most fundamental of all our concepts and laws in physics, the first law of thermodynamics, increasing gravitational energy turns into heat. We experience it in action every day—for example, when we refill a bicycle tire with air, the pump heats up. The excess heat is radiated away. Within the clouds, baryons dissipate and cool. The gas fragments into stars. One important prediction is that the merging of cloudy substructures induces compression waves in the gas that are expected to enhance the formation of stars. We should see outbursts of star formation, especially far away, from galaxy mergers. As we look back in the Universe with our largest telescopes, we also expect to see the number of smaller galaxies successively increase relative to the largest galaxies. We should observe the star formation rate being progressively enhanced as we look back in time.

7. Although closer than Andromeda, our galaxy is more difficult to survey because a three-dimensional map has to be constructed that takes account of our noncentral location. Modern techniques use precision velocity measurements made with interstellar hydrogen emitters. The results demonstrate that the rotation speeds of stars in our galaxy stay constant out to a distance of 200,000 light-years from the center. The stars mostly extend to just 10,000 light-years away from the center and slowly peter out at longer distances. Dark matter gravity dominates and helps to boost fluctuation growth at early times.

The cold and weakly interacting dark matter formed dark halos surrounding the galaxies. Harnessed by the dark matter, the gas clouds cooled and contracted. Thousands of gas clouds assembled and contracted to form galaxies that are embedded in dark halos.

Once stars form, their radiation warms and even accelerates the gas clouds. Clouds merge, and structure develops. The Universe soon becomes more complex. It is a combination of hot diffuse gas, cold interstellar clouds, stars, and dark matter. The dwarf galaxies are the visible survivors, but there once were far more whose assembly left relics behind. There should be many more relics than observed dwarfs. We have been able to observe some of the predicted leftover structures in the Milky Way using stars as tracers. This breakthrough came about with the GAIA space telescope. GAIA consists of a pair of small telescopes linked together to provide precise measurements of positions and motions of a billion stars in the Milky Way.

8. The dark matter that surrounds our galaxy forms a giant halo. For many years, astronomers assumed that the halo was a huge monolithic blob of dark matter. There have since been many attempts to detect the dark matter directly. To this day, we have no idea of the nature of the dark matter, just that it is a source of gravity in the outer galaxy, its density is very low, and it is completely transparent.

Dwarf galaxies are initially rich in gas and easily disrupted by supernova explosions. Most of the gas reservoir is ejected by the hot outflows. Star formation is halted. The final product, lacking the usual numbers of stars, may often be too dim to be easily detectable. Searches for ultrafaint dwarf galaxies around the Milky Way have had some success. Searching for faint ghostly relics of galaxies, astronomers have reported some dozens of candidates. These are seen essentially as associations of old stars in the halo. They are discovered by counting faint stars and looking for slight excesses over the background. Such dim star piles are believed to be the relic survivors from when our galaxy formed, some 10 billion years ago.

The violent events that we characterize as supernovae follow the deaths of the first generation of massive stars. They have a dramatic effect on the remaining gas reservoir in the first clouds. There can be no subsequent star formation, at least not until more cold gas is supplied by accretion from the surroundings. This is rather unlikely, as the dwarfs are orbiting their parent galaxy at high velocity. Little cold gas remains in the halo after the Milky Way forms.

Many old dwarf galaxies have been observed, although the numbers fall short. Their frequency is still below the predicted numbers. Something must have happened to them. We believe that it was a combination of gas stripping, resulting in inefficient star formation, and tidal disruption, especially for the dwarfs that pass close to the dense inner galaxy and disk.

9. Astronomers survey the relative amounts of iron throughout the galaxy. They find that the inner regions of our galaxy are progressively more enriched than are the outer regions. The galaxy formed inside out. The enriched debris from the first generations of stars is captured into the central regions of the galaxy. From there it is

dispersed throughout the galaxy into interstellar clouds. These clouds successively form new generations of stars. The Sun formed about halfway through this process, some 4.6 billion years ago.

Chapter 3. Robots and Humans

1. Lunar water was first discovered by an Indian probe and confirmed by a NASA spacecraft. What was first discovered was the hydroxyl radical molecule, which is a good trace of water ice, as it forms when ice is dissociated by ultraviolet light; it is not, however, a direct proof of the presence of water. In principle, the oxygen in lunar regolith can combine with the solar protons that bombard the moon to form hydroxyls. A recent mapping of the Moon focused on the second-largest crater visible from Earth, Thiome, which is 140 miles across and two miles deep. The mapping was accomplished in 2020 with the SOFIA infrared telescope. SOFIA is built into a modified Boeing 747 that flies at 45,000 feet, an altitude that is above almost all terrestrial infrared emissions that would cast an unacceptably bright background glow. SOFIA looks at faint cosmic infrared sources such as distant galaxies, but it has also observed the Moon. The telescope detected direct evidence for water molecules in sunlit lunar craters. More importantly, the discovery strongly suggests that there may be large water deposits to be discovered in more shielded regions.

2. We will begin our lunar exploration strategy on the Earth. Key mission elements will be built as terrestrial prototypes, including astronaut training facilities and manufacturing modules using local resources, and tested in analog environments. Test campaigns will use lunar rovers and infrastructure elements adapted for the preparation of the necessary terrestrial communication infrastructures and human interfaces. Robotic functionalities will allow autonomous deployment and maintenance of scientific instruments to enable inspection of the environment through precise three-dimensional mapping. Robots will even select the optimal sites for telescopes.

3. To achieve the needed degree of coherence in the radio signals received from the dark ages we use the principle of interference. The idea is that crests of light or radio waves add up to produce bigger crests, and troughs add up to produce deeper troughs, but any mismatch between the troughs weakens the signal. Introducing tiny time delays helps us maximize the signal and allows reinforcement and coherent addition of the many antenna signals. Each separate antenna receives a slightly different phase or sequence of crests of waves. By arranging to collect signals that do not cancel each other out, using precise clock timing, the signals from many radio dipole antennae can be combined electronically in order to generate an image. Collecting all the phases with an electronic signal correlator enables the image to be restored. Waves now reinforce each other. Otherwise, it just amounts to a jumble of data points.

The major obstacle will be the radio foregrounds that emanate from our galaxy and from many other galaxies throughout the Universe. The collective glow is millions of times brighter than the feeble signal for which we are searching. The radio foregrounds come from groupings of many radio sources. We need to subtract these from the detected signal in order to be able to search for the faint residual signal from the dark ages. We are seeking the intrinsically diffuse glow from many clouds. Fortunately, there are ways to do this, as the elusive signal has a unique structure in frequency and in space, but it will not be easy. Examples of radio interferometers include the VLA in Socorro, New Mexico, and ALMA in Atacama, Chile. These telescopes usually operate at higher radio frequencies. In the low-frequency range, closer to what will be needed for a dark ages probe, the largest interferometer so far is LOFAR, which has some 20,000 antennae scattered throughout Europe. Current plans include construction of the low-radio-frequency SKA, which will observe at radio frequencies as low as 30 megahertz and with up to one million antennae in the Australian desert, as well as with a higher-frequency counterpart in South Africa.

4. The technical details of lunar telescopes are enormously complex, and they will also be in competition with other demands for lunar science projects. Constructing them on a polar site is crucial in order to obtain a permanently dark and cold platform. There are environmental issues to be resolved, such as overcoming the abrasive and ubiquitous nature of lunar dust and managing its effects on the mirror degradation, lifetime, and observational quality of the telescopes. The infrastructure requires excavation, construction, deployment, and assembly. We will need to develop foundations, landing pads, roads, cable trenches, transportation, and means of heavy lifting. Robotic assembly will be critical. Maintenance and storage that account for climatically extreme conditions will be essential. None of these challenges, however, are insurmountable. Similar infrastructure will enable other commercially driven projects.

Chapter 4. Tuning in to Our Origins

1. Eventually the overdensities form clouds, once the ordinary matter is mostly atomic. Gravity only becomes effective in enhancing density fluctuations once the Universe has cooled sufficiently for the pressure of the radiation to drop relative to that of the matter. Radiation exerts a friction as the baryons are scattered. The baryons are effectively braked as clouds contract. Thanks to the pull of the dark matter, slightly overdense spots get denser. Gravity drags in more stuff, both dark matter and ordinary matter. There are as many underdense spots as overdense spots. The underdense patches get emptier and emptier as gravity operates. A lack of matter acts like antigravity. It is as though the rich get richer, and the poor get poorer. The first clouds form, and so do the first voids. But it is the clouds that are destined to aggregate with

each other under mutual gravitational attraction to form the galaxies. These are the culmination of cosmological growth.

At first, the clouds cannot cool further. They are not cold enough for hydrogen atoms to condense and fragment into stars. We have seen that there were a few unattached electrons, about one in 1,000 in fact. The free electrons act as a catalyst. They attach to hydrogen atoms to form charged molecules. These in turn unite with protons to form hydrogen molecules. Very soon about one particle in 1,000 is a hydrogen molecule. Hydrogen molecules can lose energy much more efficiently than hydrogen atoms. So the gas clouds cool dramatically, to about 1,000 degrees Kelvin. Once the first stars appear, the dark ages are over.

2. The Big Bang left its elusive fingerprints on the cosmos, clues to how the Universe began—but the mystery remains unsolved. The scientific consensus is that all we see in the visible Universe emerged from a period of immensely rapid expansion that occurred some trillionth of a trillionth of a second after the Big Bang. These events left behind an electromagnetic radiation fossil glow that is detectable to this day, 13.7 billion years later. This cosmic relic is the cosmic microwave background. It has provided an image of the Universe from long before galaxies formed, and it has shed light on the seed fluctuations from which galaxies formed. But it lacks the sensitivity to glimpse the beginning. The cosmic microwave background has millions of pixels on the sky, but the information content is lacking.

There is only one way to improve cosmological precision another hundredfold. We need to go beyond galaxies to their building blocks, the millions of gas cloud precursors of each galaxy that are detectable only by searching in the dark ages. Being able to target them would give us an increase of a hundredfold or more over what can be done with galaxies.

3. A collision with another hydrogen atom at very low energy suffices to bounce the electron spin from being antiparallel to the central proton to parallel. This requires absorbing a tiny amount of energy from the collision. We observe this tiny deficit as a 21-centimeter shadow against the cosmic background radiation. The increment in energy is radiated away by photons of the frequency 1420,405751.7667 megahertz, or wavelength 21.1061140542 centimeters. That's where the shadow is. The electron spin flips to the level of the higher parallel spin. The number of atoms in the slightly higher-energy parallel state is also sensitive to the temperature of the cloud. So we can even measure the cloud temperature as well as the number of cold atoms of hydrogen.

4. Before the first stars formed, obscurity prevailed. The dark ages are the period after the last scatterings of the cosmic microwave background. This cast a veil over the preceding 380,000 years since the Big Bang. The epoch of the first stars, about one billion years ago, likewise obscures the dark ages. We need to search in between the epoch of last scatterings of photons and the formation of the first stars. During

this completely unexplored period, the seeds for structure were laid down in gaseous form. The Universe was completely dark.

During this phase, we need to detect the hydrogen atoms between a redshift of 30, or 100 million years after the beginning, and 80, or 2.4 million years later. Over this window into the past, the hydrogen 21-centimeter wavelength is stretched to 6.3 meters and 16.8 meters. The hydrogen 21-centimeter line frequency is lowered to 67 megahertz and 18 megahertz. The sweet spot is at redshift 50, where the clouds are at their relative coldest compared to the background light, for an observed wavelength of 10.5 meters and an observed frequency of 28 megahertz. This sweet spot allows the determination of a precise redshift. Of course, it will be partially blurred by the redshifting of frequencies due to the expansion of the Universe. However, armed with this precision, we can slice up the Universe into layers as we look back into the past. It is tomography—the same technique that is used in medical imaging with X-rays—that allows us to do this. Each frequency corresponds to a look-back time. We bin in frequencies. So now there is a series of slices in time or, equivalently, slices in distance.

5. We run out of pixels with any cosmic microwave background sky map. Counting galaxies is better. Planned galaxy survey projects should yield some ten times more precision for cosmology than we have now. We expect 100 million independent bits of information on the sky. So over the next decade we will achieve a precision of one part in 1,000 in cosmology. This is ten times better than the precision we presently have from the cosmic microwave background. Even that is not enough! The necessary combination of a wide field of view and a large aperture for any feasible terrestrial telescope is impossible to achieve, even though the largest galaxy surveys are capable of providing so much more precision than the level attainable with the cosmic microwave background.

6. There is a natural limit to what can be observed since galaxies have finite size. We want to avoid overlapping images, which creates confusion. But the most distant galaxies are barely resolvable. We could still use billions of pixels before running into any problems of overcrowding. We need more than just images, however, to learn about the physics of the distant galaxies. The standard procedure in astronomy is to divide the light up into many wavelengths and obtain a spectrum. That provides more information, but there is a cost to this procedure. Each telescope aperture has a certain light-gathering power. Slicing the light up by wavelength in order to obtain a spectrum restricts us to studying only brighter galaxies. There are so many photons. We will learn much about galaxies, such as how they age and their chemical composition, but we will not be able to survey space to the depth needed to really probe the origin of the Universe.

Where to go next? Not even the next-generation, ultra-large-scale galaxy surveys provide enough accuracy. The pioneering Sloan Digital Sky Survey, run by over a dozen collaborating institutions, did wonders by mapping a million galaxies over one-third of the sky. It revealed the immense power of statistics. Only with a truly large sample could we begin to probe the subtle varieties of galaxy evolution in the nearby Universe. A large telescope was not needed, but the telescope had to be dedicated to the survey. The 2.5-meter telescope was custom-built for the survey at Apache Point in New Mexico that began in 2000. Yet a million galaxies is not enough.

The successor Dark Energy Survey of galaxies used a larger four-meter telescope on Cerro Tololo Mountain in Chile to look more deeply into the Universe. It targeted hundreds of millions of galaxies. Right now that's where we are: counting galaxies with precision measurements. But even these numbers may not suffice if we are to test the inflationary hypothesis. On the ground, the 8.4-meter Rubin Telescope (dedicated to pioneering extragalactic astronomer Vera Rubin) will take us even deeper into the Universe. It is located on the summit of Cerro Pachon, at 2,680 meters above sea level in the high Atacama Desert in Chile. The Large Synoptic Survey is a key project that will provide the ultimate ground-based survey, to begin in 2023. The Rubin Telescope will target 10 billion galaxies.

New powerful telescopes will peer further into the earliest moments, most notably the next generation of space telescopes. One is the 6.5-meter James Webb Space Telescope; this infrared telescope has been launched by NASA and the ESA into an orbit one million miles from Earth, where the gravitational pulls of Earth and the Sun are in balance. It's an ideal stable point at which to place a space telescope. The launch occurred on December 25, 2021, from the European spaceport in Kourou, French Guiana, via an Ariane 5 rocket. It arrived a month later at its final destination, and the next four months were used to align and focus the segmented telescope mirrors. The infrared space telescope will be able to take exquisite images in the near-infrared of the centers of galaxies. It will search for traces of the first stars in the Universe, elucidate the growth and evolution of quasars, and study exoplanets.

One complementary project in space is the ESA's 1.2-meter Euclid Space Telescope. Euclid will be based at the L2 Lagrange point, a point of neutral gravity. Here terrestrial and solar gravity pulls cancel out to give a stable parking point for a telescope located one million miles from Earth. There are estimated to be of-order-10 billion galaxies in a typical modern deep survey such as that to be begun in 2023 when Euclid is launched. Euclid will focus on the epoch a few billion years ago when the Universe was about half its present size. Each galaxy is different, and a minimum number, at least 100 galaxies per bit of information, is needed to have a reasonably independent sample of independent points. So we infer that within the large-scale structure of the galaxy distribution there are about 100 million independent samples. Each sample is one mode of information. We can equally think of these patches as giant pixels.

Another major telescope, to be launched by NASA in 2027, will be the Nancy Roman Space Telescope, a 2.4-meter telescope characterized by its large field of view with visible and near-infrared light cameras. This is largely aimed at advancing the frontiers of cosmology, whereas the small field of view of the James Webb Space Telescope optimizes its science for individual infrared sources from exoplanets to the first galaxies. The Roman telescope will specialize in surveys that greatly extend the reach of the Euclid telescope.

7. Remarkably, our search takes us deep into the Universe to the moment where direct imaging fails. This is before the moment where we actually see the microwave background radiation from when it was last scattered by electrons. The last scattering occurred some 380,000 years after the Big Bang. Such observations determine the deepest image we can ever take—at the moment when the universe first went neutral and the atomic phase of hydrogen dominance began in earnest. That is where we seek the ultimate signal. That's the dark ages.

8. The bell curve applies to most randomly sampled sets of data. As we sample more and more data, the distribution settles into a bell-shaped curve determined by its mean value and the width of the distribution. It is used widely in statistics to characterize the rarity of deviations from the mean value. For example, in a bell curve, 99 percent of the data lie within three standard deviations of the mean.

In cosmology, we are interested in very rare deviations from the mean. If these are found, they will tell us something new about the initial conditions from which the data were generated. For instance, many parameters drive models of star formation. Inclusion of the resulting complexity leads to a predicted bell-shaped curve for the distribution of the masses of the newly formed stars. What has to be deduced—and this is much more difficult—is the spread of the mass distribution. There should not be vast numbers of giant stars unless the initial theory is biased in this direction. Likewise, there should not be vast numbers of tiny stars. Of course, nature might not agree, and this might even be the case under special circumstances. All we can do is discuss the most likely options and report on the observations.

Here is what we can calculate. What is the mass of the most massive star in a certain region of the galaxy? This calculation is useful if we want to count the number of black holes. Now we are measuring black holes when they merge with other black holes. We detect flashes of gravity waves. Or what is the smallest mass of a star? Again, this is a useful exercise if we want to estimate the contribution of dark stars to the mass budget of the galaxy. We have a firm limit on the amount of baryonic matter in the Universe. And we infer that baryons fall far below the abundance level needed to explain the dark matter in the Universe. But perhaps there were dark baryons that made dark stars. This line of reasoning may seem convoluted, but physics is capable of pointing us in this direction as we desperately try to identify the nature of dark matter. Long ago, clouds of dark baryons could have formed dark stars, which we cannot see directly.

It is tough to test this hypothesis. Our detailed data on stellar masses are all local: our studies are based on measurements of star-forming regions in the Milky Way and in a few nearby galaxies. We are trying to decipher the distant Universe. We have no guarantee that the processes that govern nearby star formation are the same in remote corners of the cosmos. Indeed, we have no way of acquiring evidence for the presence of dark stars even if they exist. This makes it questionable to apply bell curves and Gaussian statistics to distant star formation. We simply cannot be sure that the same physics applies everywhere that we point our telescopes. Looking for remote indicators of non-Gaussianities is a way of establishing whether something unprecedented is occurring far away. We need to keep an open mind. For our local patch of the Universe, however, it is just fine to use bell curves.

Chapter 5. The First Months of Creation

1. Going to truly high energies requires a new understanding of both particle physics and cosmology, beyond the classical idea of Bohr and Einstein. For comparison, the proton has a mass of one giga-electron volt. The world's most powerful particle collider, the LHC, smashes particles together at energies of 14 tera-electron volts. That is how we understand that protons are not fundamental particles but are made of quarks. Achievement of much higher energies was feasible in the Big Bang. But of course this only happened once. The fundamental forces of nature are united at an energy of 10 trillion tera-electron volts.

The grand unification theory, known loosely as GUT, spells out the magic of the inflationary moment. The horizon is the distance any particle traveling at light speed has moved since beginning. Causal contact is made on superhorizon scales. These are scales much larger than that of the distance that light has traveled since the beginning of the Universe. There is far more matter outside our current horizon that once entered briefly into contact.

Inflation was conceptualized in 1981 by particle physicists Alan Guth, Andrei Linde, Paul Steinhardt, and Andreas Albrecht. Working mostly independently, these scientists argued that the introduction of a new field would accelerate space just when unification of the fundamental forces was breaking down as the Universe expanded and cooled. We can visualize the moment of inflation as a phase transition. Just as when a lake is frozen, energy in the new field is liberated. This is a change of phase. Ice and water coexist but represent distinct phases. In the cosmological situation the phase also changes. We begin with a Universe where the new field, along with vacuum quantum fluctuations, contributes negative energy. This is released when the energy of the new field first becomes dominant. The effect of pumping up the vacuum energy accelerates the expansion, as Lemaitre found long ago. Space accelerates at an exponential rate. This is inflation. The quantum phase change injects energy into the expansion of the Universe. In response, the Universe accelerates for a brief instant. All is over when the separation of phases is completed.

Inflation ends when the new field decays to one where the fundamental nuclear forces prevail and there is no longer any vacuum energy. Normal expansion resumes. The horizon has returned to its normal extent determined by the time elapsed since the Big Bang.

The density fluctuations that seed structure formation originate in the quantum theory. It is a prediction on tiny scales, corresponding to the horizon very soon after the Planck instant of time. The fluctuations initially are infinitesimal. But space briefly undergoes a stupendous amount of inflationary growth, and then it goes back to where things were before inflation—with one key difference. The fluctuations become greatly extended during inflation. As the Universe resumes its usual expansion, the horizon continues to grow again from its infinitesimal scale, as determined by the light travel time. As the Universe ages, fluctuations continue to enter the expanding horizon, on larger and larger scales.

2. The data on the expansion rate of the Universe divide into distinct sets that prefer a low expansion rate in the early Universe and a higher value today. Generally, the data from nearby galaxies prefer a high value, or a younger Universe, and the distant data from the early Universe favor a lower value, or an older Universe. If both data sets are correct, then our theory of the expanding Universe needs revision. The implications of the dispute are important. The inferred age of the Universe is uncertain by two billion years. This depends on which data combinations one takes. A lot can happen in two billion years, and the discrepancy is worrisome for cosmologists. The result may also be due to as yet unknown errors in the data. For example, the estimates of measurement errors (around 2 percent or less) are far less than the 10 percent discordance between rival groups that use different techniques. Astronomers are trying hard to obtain improved and more precise data.

An intriguing possibility, of course, is that the discrepancy may be an indication of new but unknown physics, yet to be discovered. This after all is what motivated Einstein's search for a new theory of gravitation. As of now, however, astronomers have doubts about the measurement divergence. For example, in order to derive the rate of expansion of the Universe a distance ladder has to be constructed connecting local distance calibrators, the best known being Cepheid variable stars, to distant indicators, such as supernovae. However, use of the brightest stars in a nearby galaxy does not agree with other calibrators. The physics of stars is well understood, so this is puzzling. And this uncertainty casts a shadow on the early Universe comparison. Hopefully, the next generation of cosmology telescopes will be able to clarify these discrepancies. We will learn whether the Universe is expanding forever or whether we face a big crunch in the distant future. We will learn whether Einstein's cosmological constant is indeed constant. We will learn whether the Universe will continue to accelerate in the future. We will learn whether new physics is needed to account for the measured expansion rate of the Universe.

3. Neutrino particles are produced directly in nuclear reactions. They spin—indeed, this is how they were first predicted. Spin is a fundamental property of elementary particles. As new particles are produced in complex interactions such as decays, spin is a quantity that must be conserved. For every type of particle that spins to the right, another particle must spin to the left. Spin has to be conserved in elementary particle nuclear interactions. When the neutron was discovered, the existence of the neutrino was inferred. Eventually it was detected. Now we are able to routinely detect high-energy neutrinos from the Sun. This measurement confirms that deep inside the Sun the energy that powers our nearest star comes from thermonuclear reactions. The Sun beams neutrinos toward us. These are generated in its core, where hydrogen is burned into helium by thermonuclear reactions. They are energetic and essentially massless particles. The high energies facilitate their detection in terrestrial experiments. The neutrinos scatter in giant tanks of purified water where they produce energetic electrons and muons. These move slightly faster than the local speed of light in the water and produce a flash of blue light that is effectively an optical shock wave, analogous to a supersonic boom from an aircraft exceeding the speed of sound in air. Many phototubes surround the water tank and detect the rare light flashes. This is how solar neutrinos are observed—in what is effectively a neutrino telescope. In the first second, the Big Bang produces photons, protons and electrons, and neutrinos. We detect the photons. Stars form from clouds of hydrogen. The Big Bang theory predicts that there should be an isotropic glow of neutrinos in the sky. So far it has not been possible to detect them because their energies are too low.

Helium was first discovered in the early days of spectroscopy in the nineteenth century, when astronomers first began to probe the composition of the stars. The sun shines because it is fusing hydrogen into helium by thermonuclear reactions in its core. The four particles of a nucleus of helium, two protons and two neutrons, weigh slightly less than four hydrogen nuclei. It is this mass difference of 0.7 percent of the original mass that is converted into energy via Einstein's famous equation, $E = mc^2$, that converts nuclear mass into energy. We exist on Earth because of the release of solar thermonuclear energy. We measure helium in many different stars and in interstellar gas, from which stars are born. Everywhere we look, near and far, there seems to be a similar amount of helium once we correct for the small amount that stars have synthesized.

Helium must have been produced long before there were stars. There is simply far too much helium around us in our galaxy to have been synthesized in stars. We even see helium in distant galaxies and in the gas between the galaxies. Helium preceded all of the stars. At an epoch of a second, the Big Bang consisted of protons, electrons, positrons, and neutrons in a sea of photons and neutrinos. Indeed, much earlier, there were no protons or neutrons, just their constituents, quarks. After a minute or so, the temperature cools sufficiently that a proton captures a neutron. This is a nucleus of heavy hydrogen, or deuterium. This in turn captures another neutron to form tritium, a short-lived isotope of hydrogen. More reactions follow, and the net result is the formation of helium.

The two protons and two neutrons contained in the nucleus of each helium is a symmetric combination that makes it very stable. Helium is the culmination of early Universe synthesis of nuclei. The typical energy of a photon in the cosmic microwave background today is about one-thousandth of an electron volt. That corresponds to a blackbody temperature of about 3 degrees Kelvin. Once the fossil radiation was much hotter. One second after the beginning the temperature was 10 billions of degrees. That is about one million electron volts of energy per photon. Remarkably, we have fossils from this moment in time. These are the helium nuclei, the second most abundant element in the Universe.

4. The cosmic microwave background radiation has the most perfect blackbody spectrum ever observed in nature. The principal investigator of the microwave background spectrum experiment was NASA scientist John Mather, who received the 2006 Nobel Prize in Physics for his work.

Chapter 6. Our Violent Past

1. X-ray emission is an inevitable consequence of accretion from a massive star onto its compact companion. We infer the masses of both objects from the measured orbital speeds and separation. In many cases, we infer the presence of a black hole. For a review, see Haardt et al., "Astrophysical Black Holes."

2. One intriguing historical footnote is that the existence of massive self-gravitating objects so compact that light cannot escape was conjectured long before Einstein's discovery. This conjecture was based on Newton's theory of gravity, developed some two and a half centuries earlier. The English natural philosopher John Michell wrote in 1783:

> If the semi-diameter of a sphere of the same density as the Sun were to exceed that of the Sun in the proportion of 500 to 1, a body falling from an infinite height towards it would have acquired at its surface greater velocity than that of light, and consequently supposing light to be attracted by the same force in proportion to its vis inertiae, with other bodies, all light emitted from such a body would be made to return towards it by its own proper gravity.

Michell went on to predict the existence of dark stars. He even predicted the existence of binary systems combining a dark star and an ordinary star. A decade later, French mathematician Pierre-Simon Laplace independently suggested the same idea in *Exposition du Système du Monde*, published in 1796. Michell and Laplace were far ahead of their time. Dark stars, reincarnated as black holes, would only resurface some two centuries later.

3. A neutron star has a density of some 10^{15} grams per cubic centimeter. If a neutron star is too massive, it collapses to a black hole. The maximum mass of a neutron star is calculated to be about three solar masses. The parent star loses matter by driving outflows and winds as it evolves. The end point is a neutron star. If the final object weighs more than two solar masses, a black hole forms. There are many black holes of stellar origin. Black hole binaries with ordinary stars are the most prominent manifestation of stellar black holes. We detect them via the X-ray emission generated when material outflows from the companion star onto the black hole. This happens when the companion star begins to exhaust its nuclear fuel as it ages and swells up to become a supergiant star.

4. The effect produced by the passing gravity waves is tiny because such gravity wave events are very rare. So they must occur far away. We would have to monitor many remote galaxies to catch a rare black hole merger. Typically, we would survey galaxies that are thousands of millions of light-years away. Detection of gravity waves from a merger event requires measuring changes in the length of a measuring rod due to the distortion caused by the varying gravity field to one part in a billion trillion.

The varying gravity field is caused as gravity waves from a distant black hole merger pass by our gravity wave detector. The experiment designed to measure gravity waves is a laser interferometer. There are two interferometers in the United States, one in Washington State and one in Louisiana, and they simultaneously saw the same signal. The measuring rod is equivalent to the length of the laser beam. It required forty years of experimental development and improvement on the initial design for this experiment to succeed. The name of the gravity wave telescope is LIGO, which stands for laser interferometer gravitational wave observatory.

LIGO uses the technique of interferometry to give precision distances. It has two perpendicular tunnels, each four kilometers long, that terminate in mirrors that reflect the laser light. The beams are reflected hundreds of times, so that the effective beam length or arm of the interferometer is hundreds of kilometers. Making use of the principle that waves have crests and troughs, an interferometer aligns the wave crests from the two beams. Distances can thus be measured to an accuracy as precise as the wavelength of the laser light. Normally, when the two laser beams converge in the central part of the detector, the troughs and peaks of the waves cancel out. The incidence of the gravity wave is slightly out of phase for the two perpendicular beams. The crests and troughs of the gravity waves are distinguished by monitoring the laser beams. When the beams are added together, a pattern of interference is produced, enabling the crests to be aligned and the signal to be reinforced. The combined beams give the gravity wave signal.

The more massive the black hole, the larger it is, that is, the larger its Schwarzschild radius is. A larger black hole has a longer light crossing time. The corresponding gravity wave traversal time is also increased. This results in a lower gravity wave frequency. Hence we can infer the black hole masses from the final chirp as the black holes merge. Before then, the black holes inspiral toward each other in a series of ever-faster orbits. We can infer the distance to the merger event because the expansion of the Universe imparts a lowering in frequency of the waves emitted due to the redshift effect on the wave form of the signal.

Massive stars currently are seen to lose a lot of their mass to stellar winds as they age. The winds are largely driven as intense radiation from the hot interior penetrates a fog of opacity in the atmosphere of the star. The momentum of the radiation drives the wind. But the fog is greatly reduced in the near-absence of heavy elements heavier than helium. Massive stars in the distant past had greatly suppressed winds, and it is these that are the progenitors of the observed massive black holes. Currently, massive stars lose most of their initial mass and form lower-mass black holes. We have learned that most of the black hole mergers occurred long ago at an epoch in the Universe when the parent stars were deficient in heavy elements. The chemical evolution of the Universe takes billions of years. Enrichment to solar values takes about five billion years. The lack of heavy elements in the early stars reduces mass loss and allows the formation of more massive black holes than are seen locally.

The first detection of a merger event corresponded to a pair of black holes a billion light-years away. Because merger events are rare and also very short-lived, there were surprises. The masses of the black holes were about twice what had been anticipated from observations of X-ray sources in the galaxy. There were further surprises. Astronomers had studied how black holes form as stars die. One prediction was that the most massive black hole that can form by collapse of a massive star should not exceed fifty solar masses. More massive stars self-destruct as they die. Another prediction was that the smallest black hole should be at least two solar masses. Massive stars that are less than about twenty solar masses form neutron stars when they die, and neutron stars have masses up to about two solar masses. Observations did not completely agree, however, with the theorists' expectations. It seems that the most massive black holes observed by LIGO are more massive than expected, and also that the smallest detected black holes have less mass than the astronomers had predicted. The numbers are relatively small. Future observations will be really useful, but for a significant advance, we must wait for the next generation of gravity wave telescopes to become available in the 2030s.

It is not straightforward to interpret the current data, and there are various hypotheses that could fill in the gaps in our models. Nevertheless, it is always exciting to be surprised. The new discoveries fuel new insights and new speculations. The observations will inspire more refined theories and acquisition of more data. This is how physics works. For the moment we have confirmed known physics. Gravity wave emission is initially very weak when the black holes are far apart. As the black holes gradually spiral together over many orbits, the gravity wave signal intensifies and rises toward higher frequencies. This is how we observe the evolution of the merging event. We have drawn two remarkable conclusions from LIGO. Black holes exist, and general relativity is confirmed. That is, Einstein's theory of gravity works.

5. The quasars are a far more powerful cosmic analog of the massive black hole in our galaxy. They are hundreds to thousands of times more massive and are mostly found in the far reaches of the Universe. The center of our galaxy resembles a scaled-down version of a quasar that was last active millions of years ago. The luminosity of the quasar is produced in a highly compact region. All happens within the central accretion disk around the black hole. Yet quasar luminosities exceed the cumulative emission of tens of billions of stars in the host galaxy. Indeed, quasars are so bright that all that is seen initially is a bright starlike object whose emission is totally different from that of any known star. A supermassive black hole was finally directly detected in 2019 by the Event Horizon Telescope. It is located at the center of the nearest giant elliptical galaxy to our Milky Way, some 50 million light-years away. Even the Milky Way has a central massive black hole. The 2020 Nobel Prize in Physics was awarded to astronomers Reinhard Genzel and Andrea Ghez for the discovery of a massive black hole at the center of our galaxy. The numbers of luminous quasars increase as we look back in time. Quasars were especially prolific when the Universe was young, a few billion years ago. This corresponds to the epoch when many quasars formed.

6. Prior to 2016, conjecture and inference, based on sound, albeit indirect, physical reasoning, reinforced our belief in the existence of black holes. All has changed since then. First was the direct detection of black holes produced by the deaths of massive stars by the LIGO and VIRGO gravity wave telescopes. On the supermassive black hole front, the major breakthrough was the discovery in 2019 of the supermassive black hole in a nearby galaxy, Messier 87, via its radio shadow.

Light cannot escape from a black hole. The black hole shadow was revealed by the radio image. The discovery of the shadow gave the first direct image of the horizon of the black hole, the point of no return for any space traveler. The inferred mass was a stunning six billion solar masses for the black hole centered in Messier 87.

The centers of galaxies once were the sites of the most energetic phenomena in the Universe. These have been recognized by astronomers as quasars. Such objects lurk in the cores of galaxies. They were luminous when the Universe was young and gas-rich. Current observations suggest that supermassive black holes formed at the same epoch as the first galaxies. Magnetic fields also play an important role. They are responsible for the radio emission that highlights the jets. The flywheel of the rapidly spinning massive black hole is gripped through powerful magnetic fields, thought to be omnipresent. The fields initially have a dipole pattern. Soon, complex structure develops because of the differential spin. Inner rings of material overtake outer rings, and the fields become highly turbulent and tangled. Magnetic field lines of force initially have opposite, or north and south, polarities that intersect each other. This leads to continuing field reconnection as the turbulent motions drag more and more fields together. Field reconnection provides energy, and this energy release powers the quasar.

7. There was a cooling channel for the first clouds. Cooling occurred by excitations of hydrogen atoms. But it was inefficient, requiring trace amounts of hydrogen molecules. The clouds cannot be too low in mass because, in low-mass clouds, the hydrogen atoms do not have enough energy to collide with the molecules and stimulate the cooling. This early channel for clouds to lose energy is missing. Collapse and eventual fragmentation into stars requires a sprinkling of extra coolants like carbon. This happens later in the successor generations of clouds, but in many of the first clouds cooling is not effective enough for the clouds to fragment into stars. They collapse into black holes. Some will be very massive. Supermassive black holes are mostly inactive today. But long ago they were hyperactive. They were accreting lots of gas from their surroundings. We see very massive black holes in the early Universe. They are the objects that we identify as quasars. They are visible because they are accreting so much interstellar gas. The gas heats up and glows in X-rays as the black hole is fueled.

8. Most massive elliptical galaxies today are old and red in color. Red stars are predominantly old stars. All ellipticals seem to contain massive central black holes. Because of the paucity of interstellar gas, these black holes are not being fed and are not active today. Occasionally there is a merger, most typically with a small galaxy. Some of whatever gas it contains will fall into the black hole and drive a new phase of activity.

To better understand the black hole feeding, new observations are being performed with our best telescopes. One is the ALMA radio interferometer, now taking data at high angular resolution in Chile. ALMA maps the gas reservoir that feeds the black holes and fuels the jets.

9. For many years, growing circumstantial evidence suggested the presence of a massive black hole at the center of our galaxy. A powerful explosion occurred there one million years ago. There is evidence, in the form of giant hot gas bubbles around the center, of a powerful source of energy at the galactic center. The bubbles are glowing in gamma rays and radio waves. The emission is produced by highly energetic cosmic ray electrons, which are short-lived particles that bear witness to an origin in a giant particle accelerator at the center of our galaxy. The central black hole was finally discovered after a decade-long monitoring campaign of orbiting stars.

Black hole activity can be a dramatic event. We occasionally see powerful jets of plasma that drive giant radio-emitting lobes. The radio jets are thought to be produced in the energy-producing region around the black hole by major releases of energy that arise from the rotation and winding-up of magnetic fields. The debris disk surrounding the supermassive black hole is rapidly spinning. The black hole itself is spinning at near light speed. Space is distorted by the spin of the black hole, which comes from the angular momentum of the infalling debris. We saw earlier that the black hole environment is our best bet for understanding the jets of radio emission emanating from massive black holes. The reconnection of the magnetic field releases huge amounts of energy. The energy output is initially channeled along the axis of rotation, then emerges as a jet moving at up to one-tenth of the speed of light. The jet stays collimated over thousands of light-years because of the pressure of the ambient interstellar medium.

10. Any electromagnetic light is shrouded by dense gas and dust, but gravitational waves escape freely. They are very weak and very elusive. The frequency of the waves is so low that a novel type of space experiment is needed to detect them. To catch the low frequency or long wavelengths of the gravity waves, the baselines must be much longer than anything achievable in a terrestrial gravity wave detector.

There are three gravity wave telescopes currently in operation. LIGO operates on two US sites and VIRGO on an Italian site. Although we have used them to discover black holes, for the most part formed in the death throes of massive stars, many questions remain. Some of the newly discovered black holes have more mass than was expected from our models of dying stars. Others have too little mass. Astronomers clamor for ever more powerful telescopes to sort out these issues. One major development is expected in 2037, when LISA, a trio of space satellites flying in triangular formation around the Sun, will be launched. They will bounce laser beams between mirrors mounted in them over separations of one million kilometers. This new telescope will open up the search window to much larger black holes with correspondingly larger horizons that emit gravitational waves at much lower frequencies. LISA will map the growth of supermassive black holes at the dawn of structure formation. Supermassive black holes are located in the nuclei of galaxies and power the enigmatic objects called quasars. Such mysterious objects are the most luminous steadily shining objects in the universe.

There is a missing link. Intermediate frequency studies are essential to understand the inspiraling of black holes and debris when the black holes of stellar origin formed. Such studies will reveal the pathway to black hole formation. For these we need beam lengths of thousands of kilometers to establish baselines intermediate between the laser beams of current detectors like LIGO and future detectors like LISA. The Moon can provide these because of its uniquely low seismic noise.

A similar concept is used with millisecond pulsars. These highly accurate cosmic clocks are rapidly spinning neutron stars that beam pulsating radio signals with millisecond periodicity. The arrival time of the pulses has a precision that exceeds the best terrestrial atomic clocks. The pulsars are separated in space by hundreds of light-years. Networks of pulsars in the sky are used to monitor gravitational waves at very low frequencies over tens of years. The use of lunar seismometers will open up a very different frequency range because we will be using the size of the Moon as the basic measuring rod. The measurable gravity frequencies are intermediate between terrestrial experiments, typically with baselines of several kilometers, and future space satellite experiments, with baselines of one million kilometers.

The prime target of the three LISA satellites scheduled for launch in 2034 will be supermassive black holes. The frequencies of the gravitational waves detected by LISA will correspond to the light travel time across the horizon of a supermassive black hole and will be in the range of seconds to minutes.

How did such supermassive black holes, with masses billions of times the mass of the Sun, assemble so quickly a few tens of millions of years after the Big Bang? Presumably by the merging of intermediate mass black holes, the missing link in our evolutionary record. And these in turn formed by swallowing many smaller black holes. We hope to see this growth in action. LIGO and its partner telescopes are measuring the mergers of stellar black holes. New kilometer-scale telescopes are under construction in Japan and India. Much larger baselines are feasible in space, with corresponding gravity waves of much lower frequencies. The space satellite trio of LISA will orbit the Sun, flying in formation in a triangle that is one million kilometers on each side.

One great mystery asserts itself. We see incredibly massive black holes long ago, at much the same epoch in the past as we detect the first galaxies. There must be a connection. But what is it? We are left with an uncertain feeling. Did galaxies form first? Or were they preceded by massive black holes in the centers of future galaxies? Our own galaxy is a prime example. A key question is how black holes accumulated so much mass. The crisis is extreme because quasars are found to be black holes with huge masses. Some are already present at the epoch of the formation of the first galaxies. Indeed, supermassive black holes may even precede massive galaxies.

When we ask where these masses came from, and which came first, galaxies or supermassive black holes, one possible answer is neither—perhaps it was a matter of coevolution. Biology tells us that common evolution is inevitable. For galaxies, it may be that the conditions that allowed the accumulation of the gas supply from which galaxies developed also simultaneously generated the extreme densities in the central regions that stimulated black hole formation. We don't know. Telescopes on the Moon will provide immense light-gathering power and exquisite resolution, allowing us to look into the immediate surroundings of massive black holes. Such telescopes of unprecedented aperture are our best bet for approaching this problem.

Chapter 7. Are We Alone?

1. "Invisibility cloaks" is one of many counters to Fermi's question about extraterrestrial civilizations: "Where are they?" Designing material with a variable refractive index has been shown to yield limited cloaking at optical wavelengths. Extrapolating this technology into a remote future capable of interstellar space travel suggests that we might be naive to ignore this possible solution of the Fermi paradox.

One notion of the adverse risks of advertising our existence to civilizations desperate to escape from the vicinity of a dying host star is vividly portrayed in a recent prize-winning science fiction trilogy, *The Three-Body Problem* by Chinese novelist Liu Cixin. Other writers and futurologists are less alarmist about future contacts.

This lack of success, at least so far, in encountering aliens is known as the Fermi paradox. The paradox is based on the thesis that if their civilizations exist in significant numbers, their advanced technology, achieved long before terrestrial technology developed, should have allowed them to visit the Earth. Of course, there are many counterarguments to Fermi's paradox, most notably involving the level of sophistication required to undertake interstellar travel.

2. Some 10,000 exoplanets had been detected as of 2021. About half were detected by transits seen by the Kepler Space Telescope between 2009 and 2018. Kepler looked only at stars in the region near our Sun. Many were less luminous than the Sun. The M-dwarfs are about one-third of a solar mass and one-tenth of a solar luminosity. M-dwarfs are much more numerous than G-dwarf stars like the Sun. And they have exoplanets. The red M-dwarfs and yellow G-dwarfs are easily distinguishable. The color reflects the temperature of the star's atmosphere. The more massive the star, the hotter it is and bluer in color. Our first target exoplanet around the nearest star, M-dwarf Proxima Centauri, has a mass that is 1.3 times the mass of the Earth.

3. Water has a few special characteristics. It is an excellent solvent. It is less dense as a solid than as a liquid. It is amphoteric, which means that it can become an acid or an alkaline base by donating or accepting a positive hydrogen ion. Water may be abundant across the Universe.

4. In addition to searching for light signals of climatic or biological origin, we can look for radio signals that might be generated by advanced technology. This reasoning has motivated the privately funded SETI (Search for Extra-Terrestrial Intelligence) program, based in California. A key element is the use of the world's largest radio telescopes. The world's largest radio telescope currently is FAST, a 500-meter-diameter dish with a filled spherical aperture. FAST is located in Guizhou Province in southwest China and was completed in 2016. It is built in a sinkhole in permeable rock. The receiver is supported by aerial cables mounted on towers that rise high above the dish. FAST is continuing the SETI search toward nearby stars.

5. Our current searches for exoplanets are limited by telescope aperture and angular resolution. The next major space telescope, planned for launch in the 2030s, is likely to have a six-meter aperture. Its spectroscopic spread over wavelengths from ultraviolet to infrared will enable it to observe a small sample of the nearest rocky and habitable-zone exoplanets. Given the limitations on telescope size for planned space missions, all of the targets of the flagship exoplanet mission must be within 100 light-years of the Earth. This small survey volume is a serious constraint on discovery potential. The range of the most likely stellar hosts will extend from those with the mass of the Sun to those of about one-third of a solar mass.

Such a space telescope is also limited in resolution by the interference of light waves. The diffraction of light sets the ultimate resolution limit of 1.2 times the wavelength of light divided by the diameter of the telescope mirror. The best resolution in optical light for a telescope of such moderate size is around 3 milli-arcseconds, which can resolve the radius of Jupiter at a distance of a light-year. But it is not enough to resolve Earth-like exoplanet surface features around the nearest stars. We need to do much better; we must go beyond searching for traces of atmospheric oxygen, and we need more light-gathering power. We must design a much larger telescope if we wish to have reasonable odds of detecting novel exoplanet biosignatures. The potential returns are so immense that ways will surely be found to overcome the problems faced by lunar crater-based telescopes on the Moon.

Here is the most exciting goal of all. Terrestrial astronomers are obsessed, with good reason, with the search for exoplanets. That search requires spectroscopic indicators of life-critical surface features, including oxygen-rich atmospheres and ice deposits or water oceans. We might even anticipate finding indicators of rudimentary biological activity, such as photosynthesis. Ideally, we would like to go further and ask what is the probability of rudimentary life on an exoplanet, and no doubt with detailed imaging we could pursue this question. What is completely clear is that the larger the sample, the greater the odds of obtaining a scientific return. We are currently limited by our sample sizes. We have no idea how many Earth-like exoplanets we need to observe to get a science return. A first step is choosing exoplanets that are of similar mass to Earth, that have circular orbits in the habitability zone of their host star, and that are not too hot or too cold. These characteristics are the prerequisites for life on the planet. What seems clear enough is that where the chances of life are rare, even a 15-meter telescope will not obtain a large sample. It could scrutinize a handful of the nearest habitable exoplanets. This would be a very small sample. Sample size is a choice dictated by telescope aperture. The best current estimates—made at the most simplistic level, using atmospheric water vapor as a working definition of habitability—are that at most one potentially habitable planet could be found for every 100 rocky habitable-zone exoplanets observed. Nevertheless, a 10-meter-class exoplanet-optimized space tele-scope launched within two decades would provide a marvelous pilot program for a lunar telescope.

There are estimated to be at least 100 billion exoplanets in our galaxy. Half of these might be in the habitable zone around an M-dwarf star. If so, liquid water may exist on these planets, provided that they are not too close to the host star or too far from it. But spectral resolution will be needed.

Because the oxygen lines in an exoplanet's spectrum would be slightly Doppler-shifted relative to oxygen in the Earth's atmosphere, we need the resolving power of the spectrograph to amount to something like a kilometer per second. For the nearest exoplanets, it will be possible, though challenging, to detect oxygen in the exoplan-etary atmosphere. Oxygen would not be enough to guarantee a biological signature. There are alternative pathways for atmospheric oxygen. A multimessenger approach is essential. We need several tracers of related molecules. One example is methane, a key waste product of biological activity. Detection of gaseous methane in the in-frared part of the spectrum would naturally have to follow. Another is phosphine, which is produced by microbial activity. In 2020, phosphine was detected in Venusian clouds. Venus is an unlikely place for life, and the interpretation of the phosphine evidence is under debate.

We should note that in the shorter term a few space missions will take a first stab at detecting simple signs of life. One is the Transiting Exoplanet Survey Satellite, launched in 2018. TESS has looked for planetary transits in front of nearby stars and identified many transiting super-Earths. These are exoplanets with a mass of a few times that of the Earth. Some were found in the habitable zones of red dwarf, low-mass stars, which are lower-mass versions of the Sun and about one-third of its mass. Such exoplanets will be prime targets for further near-infrared atmosphere characteriza-tion in studies by the 6.5-meter James Webb Space Telescope launched in 2021, complemented by a survey telescope, the smaller Nancy Roman Space Telescope, to be launched in 2027.

Searches for exoplanets have to overcome the problem of light contrast, because of the glow from the host stars. The experimental resolution is to shade the exoplanet from the overwhelming glare of the parent star with a star shade, an instrument launched independently and unfurled after arrival in the vicinity of the telescope. The light contrast must be reduced by one part in a billion in order to view the occulta-tion of a twin Earth exoplanet. Such a star shade is planned for launch in 2027, to complement the Roman Space Telescope.

Our future infrared space telescopes will directly image nearby super-Earths. These are planets with rocky cores and masses approaching that of Neptune. Since imaging measures the planet's reflected light, or albedo, these telescopes will probe deep into the planetary atmosphere. This will be invaluable to our efforts to further constrain atmospheric parameters. But these telescopes will not be sensitive to Earth twins. Super-Earths, so-called because of their size, are the most easily seen exoplanets. These giant gaseous planets are unlikely to harbor life. Super-Earths are detected by eclipsing a tiny part of the host star's light, and that requires much more refined subtraction of the host star's light for Earth-like planets.

Chapter 8. Survival

1. Oxford philosopher Nick Bostrom defines existential risk as a situation whose outcome would be a complete or partial catastrophe. This translates into either complete annihilation of humanity or drastic and irreversible curtailment of its potential. Asteroid impacts are a primary example. The last major asteroid impact on the Earth, and the only major one in recorded history, was the Tunguska event. This asteroid impact in Siberia a century ago had an estimated diameter of 50 meters. It destroyed thousands of square kilometers of forest. The Earth undergoes frequent smaller impacts from meteoroids. These small rocky bodies are most likely debris from an asteroid that itself underwent a major collision with another rocky body.

In the future, there is uncertainty about the time of closest approach of Asteroid Bennu because a small uncertainty about its orbital parameters translates into large uncertainty about the date of its closest approach, that is, its closest distance from Earth. Its impact is likeliest to occur between the years 2175 and 2199, according to NASA estimates.

Our best terrestrial candidate for a massive ancient asteroid impact is the Vredefort dome in South Africa, which was created by a 10-kilometer-size asteroid impact some two billion years ago. The Sudbury basin in Canada formed 1.8 billion years ago from a similar event. The oldest terrestrial impact event is associated with the Yarabubba crater in western Australia some 2.3 billion years ago. As Stephen Hawking prophetically warned in 2016, "Although the chance of a disaster to Planet Earth in a given year may be quite low, it adds up over time, and becomes a near certainty in the next thousand or 10,000 years." But he was far too pessimistic. I'd give us 100 million to one billion years.

2. The energies involved in lunar volcanic events are not uncommon with terrestrial events. Mount St. Helena erupted in Washington State in 1908 with a similar energy. Go back a century and the Krakatau volcanic explosion in Indonesia in 1883 was 10 times more powerful. Before that, in 1815, a volcanic eruption in Tambora, Indonesia, was 100 times more powerful. Crater Lake in Oregon was created around 5000 BC by an eruption with 1,000 times more power, as judged by the volume of ejecta. These giant eruptions, which are believed to occur about every 50,000 years, occur in supervolcanos. Volcanic vents of similar power must have generated the lava flows that shaped the lunar highlands.

3. Enhanced pollutants act like a greenhouse gas layer that covers the atmosphere. Factors in the resulting ocean level rise include the melting of the glaciers and the accompanying thermal expansion of the ice. Modeling of the ocean rise involves an understanding of many uncertain factors that enter into our understanding of the Earth's atmosphere. The ocean level rise is predicted to accelerate. Estimates that the oceans will rise up to 10 meters within the next 200 years are based on models with many parameters. The values of these parameters are estimated to the best of our knowledge, but inevitably the uncertainties are large.

Another major contributor to global warming is carbon burning, which also leads to other negative consequences. Here, for example, is an account of acid rain effects:

> In the Adirondack Mountains of New York the acid rain includes a mixture of sulfuric and nitric acids from the sulfur dioxide and nitrogen oxides pouring from the smokestacks of power plants, smelters, factories, and vehicle exhausts. Over 200 lakes are dead; their aquatic life gone or dwindling. And in Scandinavia acid rain has destroyed 15,000 lakes in recent years. Inevitably, the death of a lake affects other wildlife as well; fish-eating ducks, loons, otter, mink, and even birds begin to leave, because their food and shelter have been destroyed. On the ground, acid rain leaches essential nutrients from the soil—calcium, magnesium, potassium, and sodium. It prevents some seeds from germinating; it scars leaves.—Diane Soran and Danny Stillman, "An Analysis of the Alleged Kishtym Disaster" (1982)

4. Here is one eyewitness account of the likely aftermaths of nuclear testing in Russia:

> About 100 kilometres from Sverdlovsk, a highway sign warned drivers not to stop for the next 20 or 30 kilometres and to drive through at maximum speed. On both sides of the road, as far as one could see, the land was dead: no villages, no towns, only the chimneys of destroyed houses, no cultivated fields or pastures, no herds, no people . . . nothing. I saw personally . . . a large area in the vicinity of Sverdlovsk (no less than 100 to 150 sq. km and probably much more), in which any normal human activity was forbidden, people were evacuated and villages razed, evidently to prevent inhabitants from returning, there was no agriculture or live-stock raising, fishing and hunting were forbidden.—Lev Tumerman (1972)

However, the consensus is that the devastation in Sverdlovsk was more likely an environmental disaster, exacerbated by coal burning and acid rain, as discussed by Soran and Stillman. There are still serious concerns about the long-term effects of testing nuclear bombs. It is interesting to revisit the concerns raised early in the history of nuclear testing by nuclear scientists with respect to the testing of hydrogen bombs:

Nuclear scientists were seriously concerned whether detonating the first atom bomb could trigger a chain reaction. Reassured by their calculations, they went on to test atomic bombs. Current stockpiles of hydrogen bombs suffice to cause destruction of all the world's population centers, although humanity would survive a nuclear catastrophe. Shortly after the end of World War II, the scientists who developed the atomic bombs dropped on Japan tried to envision the kind of nuclear event that could lead to the destruction of not just cities, but the entire world. The verdict that scientists at the Los Alamos laboratory and test site reached in 1945 ... [was] that "it would require only in the neighborhood of 10 to 100 Supers" to put the human race in peril.—Alex Wellerstein, "Manhattan Project" (2014)

5. The heaviest quark weighs in at about 100 proton masses. The top quark is the first quark directly detected, and it adds to the known families of five lighter quarks. These constitute what is known as the standard model of particle physics. This was confirmed by the 2012 discovery of the Higgs boson. The Higgs was the missing link in the theory that enabled our understanding of the origin of the masses of fundamental particles. This discovery completed the standard model of fundamental particles.

But we need to go beyond the standard model to decipher the enormous energies available in the very early Universe. One direction is that of string theory, which postulates the existence of extra dimensions. Strings are generalizations of elementary particles in higher dimensions. They are currently hidden, being too compact, but they may open up at sufficiently high energies. One theoretical consequence is the formation of tiny black holes.

In the very early Universe, tiny black holes rapidly evaporate and have no long-term consequences. But that may change if they are produced in an ongoing particle collider experiment. Prior to the completion of the Large Hadron Collider (LHC), some critics even speculated that such tiny black holes might have catastrophic consequence for the Earth. The environmental impact of black hole formation at collider energies was actually resolved in a Hawaiian court of law, following a lawsuit. Since far higher energies than are ever achievable in a collider naturally occur as high-energy cosmic rays impacting the Earth's atmosphere, without producing any black holes, it was decided that any collider risks of black holes escaping and potentially wreaking massive destruction on the Earth were exceedingly small.

Another way tiny black holes might be produced is not limited by any human activities. There are natural particle colliders in space, the most powerful of which are millions of light-years across. Colliding with ambient gas are powerful jets that operate as natural particle accelerators and are produced by supermassive black holes in the nuclei of massive radio-emitting galaxies. The most energetic cosmic rays measured on Earth amount to about one trillion giga-electron volts. We believe that they are generated in these huge cosmic accelerators. It is possible that, very rarely, such conditions would facilitate the formation of tiny black holes. Fortunately, these natural colliders are very far away. There is no risk from them, only a possible lack of communication between scientists and the general public.

6. We have seen that the SETI (Search for Extra-Terrestrial Intelligence) program is listening for possible radio signals from alien civilizations. An interesting suggestion to make the search more efficient is to concentrate on looking in those directions in the sky where stellar mini-eclipses are found to occur. We would infer the presence of transiting solar system planets, which would be potential targets. Indeed, we on Earth would be more readily discovered by technological civilizations scanning not our planet but the Sun.

7. According to the second law of thermodynamics, some processes are irreversible; for instance, we say that entropy increases. We can look for artificial related signals in stars. A star that was in the news in 2019, KIC 8462852—more popularly called Tabby's star, after its discoverer, Tabby Boyajian—showed unusual fluctuations in brightness, sometimes dipping by as much as 2 percent for periods of just a few days, as well as a gradual decline in light, by about 1 percent, over a century. The speculation has been that these fluctuations may represent the presence of some megastructure, such as a Dyson sphere constructed around the star. A more plausible scenario is that we are observing the decline phases after a brightening that resulted from the star having swallowed an orbiting planet or asteroid.

A few strangely red but star-forming galaxies are candidates for being full of Dyson spheres. They were discovered by inspecting 100,000 infrared-emitting galaxies detected by NASA's Wide-Field Infrared Survey Explorer (WISE) telescope. Their unusual combination of high mid-infrared and low near-ultraviolet luminosities seems inconsistent with our simple expectations when we observe high rates of star formation. The ultraviolet light luminosity is dominated by young stars. These typically track the star formation rate. The infrared luminosity is dominated by the much more abundant lower-mass stars. These are seen at longer wavelengths. The intermediate wavelength excess could be due to starlight being intercepted by Dyson-like spheres. Or it could be due to the uncertain properties of the dust grains. As usual, more data are needed to sort out this signal.

Chapter 9. Internationalization

1. The United Nations Outer Space Treaty of 1967 provides the basic framework on international space law, including the following principles:

> The exploration and use of outer space shall be carried out for the benefit and in the interests of all countries and shall be the province of all mankind;

outer space shall be free for exploration and use by all States;

outer space is not subject to national appropriation by claim of sovereignty, by means of use or occupation, or by any other means;

States shall not place nuclear weapons or other weapons of mass destruction in orbit or on celestial bodies or station them in outer space in any other manner;

the Moon and other celestial bodies shall be used exclusively for peaceful purposes;

astronauts shall be regarded as the envoys of mankind;

States shall be responsible for national space activities whether carried out by governmental or non-governmental entities;

States shall be liable for damage caused by their space objects; and

States shall avoid harmful contamination of space and celestial bodies.

"As for criminal issues: partner states may exercise criminal jurisdiction over nationals of a partner state whose misconduct in orbit affects the life or safety of a national of another partner state" (from the official records of the UN General Assembly, 1,499th Plenary Meeting, December 16, 1966).

2. Five sovereign nations are currently jostling for space on the lunar surface: the United States, China, Russia, India, and Japan. The first lunar soil return sample since Apollo was obtained by China in 2020. More such missions will follow. Several missions are needed just to survey the diverse regolith in different regions.

Chapter 10. The Next Century

1. The key to understanding early events in the Universe is study of the relics that are inevitably left behind. One such relic is a sea of electromagnetic radiation coming from every direction in the sky. Released around 380,000 years after the Big Bang when the first atoms formed and the Universe was much hotter, this radiation cooled over time to microwave frequencies. Superimposed on this cosmic microwave background are patterns from scattered photons. These are vestiges of the gravity that seeded galaxies and other massive structures in the Universe. Studies from Earth-bound telescopes and orbiting satellites have mapped millions of these tiny ripples to produce precise estimates of the age of the Universe. From these ripples, we infer the rate of expansion of the Universe. We can study the relative amounts of visible matter, dark matter, and dark energy. But we need more than the ripples to advance our understanding. The key argument in this book is that we need not millions but trillions of tracers if we are ever to achieve ultimate precision in cosmology.

BIBLIOGRAPHY

Prologue: The Moon Beckons

For a historical survey of theories of the origin of the Moon, see Warren Cummings, *Evolving Theories on the Origin of the Moon* (Cham, Switzerland: Springer, 2019). For a review of theories of planet formation, see Michael Woolfson, *The Formation of the Solar System: Theories Old and New*, 2nd ed. (London: Imperial College Press, 2014). The case for lunar exploitation is eloquently made by polar cold trap pioneer Paul Spudis in *The Value of the Moon: How to Explore, Live, and Prosper in Space Using the Moon's Resources* (Washington, DC: Smithsonian Books, 2016). As first noted by the French astronomer Camille Flammarion, some of the crater rims are so high that, near the lunar poles, the peaks are bathed in almost eternal sunlight. See Flammarion's *Astronomie populaire: description générale du ciel* (Paris, 1879); see also Emerson J. Speyerer and Mark S. Robinson, "Persistently Illuminated Regions at the Lunar Poles: Ideal Sites for Future Exploration," *Icarus* 222, no. 1 (January 2013): 122–36.

Temperatures are taken from diviner radiometer on NASA's lunar orbiter mission (LRO). J.-P. Williams et al., "The Global Surface Temperatures of the Moon as Measured by the Diviner Lunar Radiometer Experiment," *Icarus* 283 (February 2017): 300–329.

SpaceX has been contracted to provide the first commercial flights, each of which will take four astronauts to the International Space Station. The company plans to launch the first private trips made in capsules by 2022. For a description of the commercial entrepreneurial approach to future developments in space exploration, see Christian Davenport, *The Space Barons: Elon Musk, Jeff Bezos, and the Quest to Colonize the Cosmos* (New York: PublicAffairs, 2019).

For current views on the giant impact origin of the Moon, see Erik Asphaug, "Impact Origin of the Moon?," *Annual Review of Earth and Planetary Sciences* 42 (2014): 551–78, https://doi.org/10.1146/annurev-earth-050212-124057.

Cixin Liu's *The Three-Body Problem* (Chongqing: Chongqing Publishing Group, 2008; New York: Tor Books, 2014) is a science fiction novel about the risks of alien communications. Such risks are echoed by the Russian philosopher and futurologist Alexey Turchin and his coauthor David Denkenberger in "Global Catastrophic Risks Connected with Extra-Terrestrial Intelligence," August 2018, https:// philpapers.org/rec/TURGCR.

For a description of historical and current views on the origins of life, see the Wikipedia entry on "Abiogenesis," updated January 22, 2022, https://en.wikipedia .org/wiki/Abiogenesis.

For further reading on lunar exploration, see the following:

Ehrenfreid, Manfred "Dutch" von, *The Artemis Lunar Program: Returning People to the Moon* (Cham, Switzerland: Springer Praxis, 2020).
Oliver Morton, *The Moon: A History for the Future* (New York: PublicAffairs, 2019).
Christopher Wanjek, *Spacefarers: How Humans Will Settle the Moon, Mars, and Beyond* (Cambridge, MA: Harvard University Press, 2020).

Chapter 1. The New Space Race

For a comprehensive history of the space race, see Roger Launius, *The Smithsonian History of Space Exploration* (Washington, DC: Smithsonian Books, 2018).

Ian Crawford and his colleagues make an interdisciplinary case for returning to the Moon in "Back to the Moon: The Scientific Rationale for Resuming Lunar Surface Exploration," *Planetary and Space Science* 74, no. 1 (December 2012): 3–14, https://ui.adsabs.harvard.edu/abs/2012P&SS ... 74.... 3C.

Many aspects of asteroid mining are covered in *Asteroids: Prospective Energy and Material Resources*, edited by Viorel Badescu (Berlin: Springer-Verlag, 2013).

Keith Veronese discusses future issues related to the rarity of rare earth elements on the Earth in *Rare: The High-Stakes Race to Satisfy Our Need for the Scarcest Metals on Earth* (New York: Prometheus, 2015).

For a recent discussion of alternative approaches to fusion energy, see Sergei V. Ryzhkov and Alexei Yu. Chirkov, *Alternative Fusion Fuels and Systems* (Boca Raton, FL: CRC Press, 2020).

Ian A. Crawford reviews the case for a moon village in "Why We Should Build a Moon Village," *Astronomy and Geophysics* 58, No. 6 (December 2017): 6.18–6.21, https://ui.adsabs.harvard.edu/abs/2017A%26G.... 58f6.18C. See also Francesco Sauro et al., "Lava Tubes on Earth, Moon, and Mars: A Review on Their Size and Morphology Revealed by Comparative Planetology," *Earth-Science Reviews* 209 (October 2020): 103288, DOI:10.1016/j.earscirev.2020.103288.

Lunar optical interferometry is discussed by Antoine Labeyrie in "Lunar Optical Interferometry and Hypertelescope for Direct Imaging at High Resolution," *Philosophical Transactions of the Royal Society A* 379, no. 2188 (January 2021): article id. 20190570, https://ui.adsabs.harvard.edu/abs/2021RSPTA.37990570L.For a more technical discussion of the potential for low-frequency radio astronomy on the Moon, see Sebastian Jester and Heino Falcke, "Science with a Lunar Low-Frequency Array: From the Dark Ages of the Universe to Nearby Exoplanets," *New Astronomy Reviews* 53 (May 2009): 1–26.

For further reading, see the following:

Jonathan Amos, "China's Chang'e-5 Mission Returns Moon Samples," *BBC News*, December 16, 2020, https://www.bbc.com/news/science-environment-55323176.

Bryan Bender, "A New Moon Race Is On. Is China Already Ahead?," *Politico*, June 13, 2019, https://www.politico.com/agenda/story/2019/06/13/china-nasa-moon -race-000897/.

Kenneth Chang, "Why Everyone Wants to Go Back to the Moon," *New York Times*, July 12, 2019, https://www.nytimes.com/2019/07/12/science/nasa-moon -apollo-artemis.html.

China National Space Administration, "China and Russia sign a Memorandum of Understanding Regarding Cooperation for the Construction of the International Lunar Research Station," March 9, 2021, http://www.cnsa.gov.cn/english /n6465652/n6465653/c6811380/content.html.

Charles Q. Choi, "Nuclear Fusion Reactor Could Be Here as Soon as 2025," Live Science, October 1, 2020, https://www.livescience.com/nuclear-fusion-reactor -sparc-2025.html.

A. J. Creely et al., "Overview of the SPARC Tokamak," *Journal of Plasma Physics* 86, no. 5 (October 2020): 865860502, https://doi.org/10.1017/S0022377820001257.

Daniel D. Durda, "Mining Near-Earth Asteroids," *Ad Astra* 18, no. 2 (2006), https:// space.nss.org/mining-near-earth-asteroids-durda/.

European Space Agency, "ESA Engineers Assess Moon Village Habitat," *Phys.org*, November 17, 2020, https://phys.org/news/2020-11-esa-moon-village-habitat.html.

Paul Glister, "A 20th Anniversary Review of Ward and Brownlee's 'Rare Earth,'" Centauri Dreams, June 26, 2020, https://www.centauri-dreams.org/2020/06/26/a -20th-anniversary-review-of-ward-and-brownlees-rare-earth/.

John M. Logsdon, "10 Presidents and NASA," *NASA 50th Magazine*, 2008, https:// www.nasa.gov/50th/50th_magazine/10presidents.html.

National Aeronautics and Space Administration, *Artemis III Science Definition Team Report*, NASA/SP-20205009602, 2020, https://www.nasa.gov/sites/default/files /atoms/files/artemis-iii-science-definition-report-12042020c.pdf.

Jake Parks, "Moon Village: Humanity's First Step toward a Lunar Colony?," *Astronomy*, May 31, 2019, https://astronomy.com/news/2019/05/moon-village -humanitys-first-step-toward-a-lunar-colony.

Monika Pronczuk, "Europe Wants to Diversify Its Pool of Astronauts," *New York Times*, February 22, 2021, https://www.nytimes.com/2021/02/22/world/europe

/women-disabled-astronauts.htmlPaul Rincon, "To the Moon and Beyond," *BBC News*, https://www.bbc.co.uk/news/extra/nkzysaP3pB/to-the-moon-and -beyond.

Paul Rincon, "Nasa Outlines Plan for First Woman on Moon by 2024," *BBC News*, September 22, 2020, https://www.bbc.com/news/science-environment-54246485.

Paul Rincon, "What Does China Want to Do on the Moon's Far Side?," *BBC News*, January 4, 2019, https://www.bbc.com/news/science-environment-46748602.

Lonnie Shekhtman, "NASA's Artemis Base Camp on the Moon Will Need Light, Water, Elevation," NASA, January 27, 2021, https://www.nasa.gov/feature/goddard /2021/nasa-s-artemis-base-camp-on-the-moon-will-need-light-water-elevation.

Mike Wall, "Presidential Visions for Space Exploration: From Ike to Biden," *Space .com*, January 20, 2021, https://www.space.com/11751-nasa-american-presidential -visions-space-exploration.html.

Wikipedia, "Artemis Accords," updated January 18, 2022, https://en.wikipedia.org/ wiki/Artemis_Accords.

Chapter 2. Digging Deep on the Moon

For reviews of the formation of the Moon, see Robin M. Canup, "Dynamics of Lunar Formation," *Annual Review of Astronomy and Astrophysics* 42, no. 1 (September 2004): 441–75; and A. E. Ringwood, *Origin of the Earth and Moon* (Berlin: Springer-Verlag, 2012).

Volker Bromm discusses the formation of the first clouds and first stars in "Formation of the First Stars," *Reports on Progress in Physics* 76 (October 13, 2013), https:// iopscience.iop.org/article/10.1088/0034-4885/76/11/112901. See also Piero Madau and Mark Dickinson, "Cosmic Star-Formation History," *Annual Review of Astronomy and Astrophysics* 52 (2014): 415–86.

Gaia results are reviewed in Amina Helmi, "Streams, Substructures, and the Early History of the Milky Way," *Annual Review of Astronomy and Astrophysics* 58 (2020): 205–56, https://doi.org/10.1146/annurev-astro-032620-021917.

For alternative theories of gravity that dispense with dark matter, see David Merritt, *A Philosophical Approach to MOND: Assessing the Milgromian Research Program in Cosmology* (New York: Cambridge University Press, 2020).

Joshua Simon reviews the status of the faintest dwarf galaxies in "The Faintest Dwarf Galaxies," *Annual Review of Astronomy and Astrophysics* 57 (2019): 375–415, https://doi.org/10.1146/annurev-astro-091918-104453.

For further reading, see the following:

Donald C. Barker, "Lunar and Off Earth Resource Drivers, Estimations, and the Development Conundrum," *Advances in Space Research* 66, no. 2 (2020): 159–377, https://doi.org/10.1016/j.asr.2020.04.001.

Jeremy Beck, "China's Helium-3 Program: A Global Game-Changer," *Space Safety Magazine*, March 19, 2016, http://www.spacesafetymagazine.com/space-on-earth/everyday-life/china-helium-3-program/.

Rebecca Boyle, "Can a Moon Base Be Safe for Astronauts?," *Scientific American*, October 22, 2020, https://www.scientificamerican.com/article/can-a-moon-base-be-safe-for-astronauts/.

Robin M. Canup and Erik Asphaug, "Origin of the Moon in a Giant Impact Near the End of the Earth's Formation," *Nature* 412 (2001): 708–12, https://www.nature.com/articles/35089010.

Felicia Chou, "NASA's SOFIA Discovers Water on Sunlit Surface of Moon," NASA release 20-105,October 26, 2020, https://www.nasa.gov/press-release/nasa-s-sofia-discovers-water-on-sunlit-surface-of-moon.

Ian A. Crawford, Katherine H. Joy, Jan H. Pasckert, and Harald Hiesinger, "The Lunar Surface as a Recorder of Astrophysical Processes," *Philosophical Transactions of the Royal Society*, A379: 20190562 (2020), https://arxiv.org/abs/2011.12744.

Francis A. Cucinotta, "Radiation Risk Acceptability and Limitations," December 21, 2010, https://three.jsc.nasa.gov/articles/AstronautRadLimitsFC.pdf.

John S. Lewis, *Mining the Sky: Untold Riches from the Asteroids, Comets, and Planets* (Reading, MA: Perseus Books, 1997).

Lunar Exploration Analysis Group (LEAG), *The Lunar Exploration Roadmap: Exploring the Moon in the 21st Century: Themes, Goals, Objectives, Investigations, and Priorities, 2016*, https://www.lpi.usra.edu/leag/roadmap/US-LER_version_1_point_3.pdf.

Lunar Surface Science Workshop, "Lunar Dust and Regolith: Outstanding Questions and Important Investigations," NASA Solar System Exploration Research Virtual Institute, August 2020 (https://lunarscience.arc.nasa.gov/lssw), https://sservi.nasa.gov/lssw/downloads/LSSW_Dust_Regolith_Report.pdf.

A. L. Melott, B. C. Thomas, M. Kachelriess, D. V. Semikoz, and A. C. Overholt, "A Supernova at 50 pc: Effects on the Earth's Atmosphere and Biota," *Astrophysical Journal* 840, no. 2 (May 2017), article id. 105, https://ui.adsabs.harvard.edu/abs/2017ApJ . . . 840..105M.

D. P. Moriarty et al., "Evidence for a Stratified Upper Mantle Preserved within the South Pole–Aitken Basin," *Journal of Geophysical Research: Planets* 121, e2020JE006589, https://doi.org/10.1029/2020JE006589.

National Academies of Sciences, Engineering, and Medicine, *Space Radiation and Astronaut Health: Managing and Communicating Cancer Risks* (Washington, DC: National Academies Press, 2021), https://doi.org/10.17226/26155.

National Academies of Sciences, Engineering, and Medicine, "NASA Should Update Astronaut Radiation Exposure Limits, Improve Communication of Cancer Risks," news release, June 24, 2021, https://www.nationalacademies.org/news

/2021/06/nasa-should-update-astronaut-radiation-exposure-limits-improve
-communication-of-cancer-risks.

National Research Council, *The Scientific Context for Exploration of the Moon*
(Washington, DC: National Academies Press, 2007), https://doi.org/10
.17226/11954.

Michele Nichols, Carleton College Department of Physics and Astronomy Good-
sell Observatory, "Mount Wilson and Palomar," June 10, 1998, https://www
.carleton.edu/goodsell/research/student-research/nichols-1998/history
/mtwilson/.

Carle Pieters, "News from Brown: Brown Scientists Announce Finding of Water on
the Moon," Brown University, September 23, 2009, https://news.brown.edu
/articles/2009/09/moonwater.

Ramin Skibba, "New Space Radiation Limits Needed for NASA Astronauts, Report Says," *Scientific American*, July 14, 2021, https://www.scientificamerican
.com/article/new-space-radiation-limits-needed-for-nasa-astronauts-report
-says/.

Michelle Starr, "The Moon's Biggest Crater Is Revealing Lunar Formation Secrets
We Never Knew," *Science Alert*, February 17, 2021, https://www.sciencealert.com
/the-moon-s-biggest-crater-is-revealing-lunar-ancient-formation-history.

Brian C. Thomas, "Photobiological Effects at Earth's surface Following a 50 pc Supernova," *Astrobiology* 18, no. 5 (May 2018): 481–90, https://ui.adsabs.harvard
.edu/abs/2018AsBio..18..481T.

Maria T. Zuber et al., "Gravity Field of the Moon from the Gravity Recovery and
Interior Laboratory (GRAIL) Mission," *Science* 339, no. 6120 (2013): 668–71,
https://www.science.org/doi/10.1126/science.1231507.

Chapter 3. Robots and Humans

For earlier studies of water on the Moon, see A. T. Basilevsky, A. M. Abdrakhimov,
and V. A. Dorofeeva, "Water and Other Volatiles on the Moon: A Review," *Solar
System Research* 46 (2012): 89–107. Additional ideas are presented in Brent Sher-
wood, "Principles for a Practical Moon Base," *Acta Astronautica* 160 (July 2019):
116–24, https://ui.adsabs.harvard.edu/abs/2019AcAau.160..116S.

For a review on the lunar dust environment, see M. Horányi et al., "The Dust
Environment of the Moon," in *New Views of the Moon 2*, proceedings of the conference held May 24–26, 2016, in Houston, TX, LPI contribution 1911, id.6005.

For a description of the Square Kilometer Array, see J. McCullin et al., "The
Square Kilometre Array Project," *Proceedings of the SPIE* 11445 (2020): id. 1144512,
https://ui.adsabs.harvard.edu/abs/2020SPIE11445E..12M/.

For further reading, see the following:

Philip T. Metzger, "Space Development and Space Science Together, an Historic Opportunity," *Space Policy* 37, pt. 2 (August 2016): 77–91, https://www.sciencedirect.com/science/article/pii/S0265964616300625.
Lawrence W. Townsend, "Space Weather on the Moon," *Physics Today* 73, no. 3 (2020): 66–67.

Chapter 4. Tuning in to Our Origins

The observational aspects of atomic hydrogen in the early Universe are reviewed by S. Furlanetto, "Cosmology at Low Frequencies: The 21 cm Transition and the High-Redshift," *Universe Physics Reports* 433 (2006): 181–301.

For an introduction to the science goals of the James Webb Space Telescope, see Jason Kalirai, "Scientific Discovery with the James Webb Space Telescope," *Contemporary Physics* 59, no. 3 (2018): 251–90, https://doi.org/10.1080/00107514.2018.1467648.

For a review of these and other future sky survey projects, see J. Anthony Tyson and Kirk D. Borne, "Future Sky Surveys: New Discovery Frontiers," in *Advances in Machine Learning and Data Mining for Astronomy*, edited by Michael J. Way et al. (Boca Raton, FL: CRC Press, 2012), 161–81.

For a review of the discovery of the acceleration of the Universe, see Adam Riess, "The Expansion of the Universe Is Faster than Expected," *Nature Reviews Physics* 2 (2020): 10–12.

For a readable introduction to the dark ages, see Avi Loeb, "The Dark Ages of the Universe," *Scientific American* 295, no. 5 (November 2006): 46–53, https://ui.adsabs.harvard.edu/abs/2006SciAm.295e..46L.

For further reading, see the following:

Anil Ananthaswamy, "Telescopes on Far Side of the Moon Could Illuminate the Cosmic Dark Ages," *Scientific American* (April 2021), https://www.scientificamerican.com/article/telescopes-on-far-side-of-the-moon-could-illuminate-the-cosmic-dark-ages1/.
Jack Burns, "Low Radio Frequency Observations from the Moon Enabled by NASA Landed Payload Missions," *Planetary Science Journal* 2, no. 44 (April 2021), https://iopscience.iop.org/article/10.3847/PSJ/abdfc3/pdf.
David R. DeBoer, Aaron R. Parsons, et al., "Hydrogen Epoch of Reionization Array (HERA)," *Publications of the Astronomical Society of the Pacific* 129, 974 (2017): 45001, arXiv:1606.07473.

Bruce Dorminey, "NASA Sites Lunar Far Side for Low-Frequency Radio Telescope," *Forbes*, August 30, 2013, https://www.forbes.com/sites/brucedorminey/2013/08/30/nasa-sites-lunar-far-side-for-low-frequency-radio-telescope/?sh=304e377b371c.

Teresa Lago, "Microsatellites Have a "Polluting Effect" and Disturb the Quality of the Sky," *World Today News*, August 22, 2021, https://www.world-today-news.com/microsatellites-have-a-polluting-effect-and-disturb-the-quality-of-the-sky-warns-astronomer-teresa-lago-observer/.

Anthony Tyson, "Cosmology Data Analysis Challenges and Opportunities in the LSST Sky Survey," *Journal of Physics: Conference Series* 1290, no. 1 (2019): article id. 012001.

Andrew Williams et al., "Analysing the Impact of Satellite Constellations and ESO's Role in Supporting the Astronomy Community," *The ESO Messenger* 184 (2021): 3–7, arXiv:2108.04005.

Chapter 5. The First Months of Creation

Inflation of the Universe is described in Alan Guth, *The Inflationary Universe* (Reading, MA: Perseus Books, 1997).

See Teresa Montaruli, "Review on Neutrino Telescopes," *Nuclear Physics B: Proceedings Supplement* 190 (2009): 101–08, https://arxiv.org/abs/0901.2661. See also Jeremy Bernstein, Lowell S. Brown, and Gerald Feinberg, "Cosmological Helium Production Simplified," *Reviews of Modern Physics* 61, no. 25 (1989), https://doi.org/10.1103/RevModPhys.61.25.

For further reading, see the following:

Augustine of Hippo, *De Civitate Dei* [413–426], book XI, chap. 6.

Dominique Lambert, "Einstein and Lemaître: Two Friends, Two Cosmologies . . . ," Interdisciplinary Encyclopedia of Religion & Science, https://inters.org/einstein-lemaitre.

Joseph Silk, *The Big Bang*, 3rd ed. (New York: Henry Holt, 2000).

Giuseppe Tanzella-Nitti, "The Pius XII–Lemaître Affair (1951–1952) on Big Bang and Creation," Interdisciplinary Encyclopedia of Religion & Science, http://inters.org/pius-xii-lemaitre.

Chapter 6. Our Violent Past

For a review of gravitational waves, see Michela Mapelli, "Astrophysics of Stellar Black Holes" (lecture summary), adapted and updated from *Proceedings of the International School of Physics "E. Fermi"*, July 3–12, 2017, Course 200 "Gravitational Waves

and Cosmology," edited by E. Coccia, N. Vittorio, and J. Silk, https://ui.adsabs
.harvard.edu/link_gateway/2018arXiv180909130M/arxiv:1809.09130.

See also Roger Blandford, "New Horizons in Black Hole Astrophysics," *Europhysics News* 52 (2021): 12–14; and Kazunori Akiyama et al., "First M87 Event Horizon Telescope Results," *Astrophysical Journal Letters* 875, no. 1 (April 2019): article id. L1, arXiv:1906.11238.

For a more technical review of radio jets, see Roger Blandford, David Meier, and Anthony Readhead, "Relativistic Jets from Active Galactic Nuclei," *Annual Review of Astronomy and Astrophysics* 57 (August 2019): 467–509.

The science of the Laser Interferometer Space Antenna (LISA) is described in Alberto Sesana, "Black Hole Science with the Laser Interferometer Space Antenna," *Frontiers in Astronomy and Space Sciences* 8 (2021): id. 7, arXiv:2105.11518.

An example of a recent proposal for a lunar gravity wave observatory can be found in Jan Harms et al., "Lunar Gravitational-Wave Antenna," arxiv preprint (October 26, 2021), https://arxiv.org/abs/2010.13726.

The quote from Kip Thorne comes from Igor D. Novikov, *Black Holes and the Universe* (Cambridge: Cambridge University Press, 1995), 1.

For further reading, see the following:

M. Bailes et al., "Gravitational-Wave Physics and Astronomy in the 2020s and 2030s," *Nature Reviews Physics* 3 (2021): 344–66, https://doi.org/10.1038/s42254-021 -00303-8.

John Baker et al., "The Laser Interferometer Space Antenna: Unveiling the Millihertz Gravitational Wave Sky," white paper submitted to 2020 Decadal Survey on Astronomy and Astrophysics, vol. 2, https://arxiv.org/abs/1907.06482.

Adrian Cho, "The Hole Truth," *Science* 371, no. 6525 (January 8, 2021): 116–19, https://science.sciencemag.org/content/371/6525/116.

Event Horizon Telescope Collaboration, "First M87 Event Horizon Telescope Results. I. The Shadow of the Supermassive Black Hole," *Astrophysical Journal Letters* 875, no. 1 (April 2019): article id. L1, arXiv:1906.11238. 2019ApJ ... 875L ... 1E.

Francesco Haardt, Vittorio Gorini, Ugo Moschella, Aldo Treves, and Monica Colpi, "Astrophysical Black Holes," *Lecture Notes in Physics* 905 (2016).

Kohei Inayosh, Eli Visbal, and Zoltán Haiman, "The Assembly of the First Massive Black Holes," *Annual Reviews of Astronomy & Astrophysics* 58 (2020): 27–97, https://arxiv.org/abs/1911.05791.

LIGO Scientific Collaboration, "Introduction to LIGO and Gravitational Waves," https://ligo.org/science.php.

Manuel Arca Sedda et al., "The Missing Link in Gravitational-Wave Astronomy: A Summary of Discoveries Waiting in the Decihertz Range," *Experimental Astronomy* 51, no. 3 (2021):1427–40, doi:10.1007/s10686-021-09713-z.

Chapter 7. Are We Alone?

For one discussion of this question, see John Gribbin, *Alone in the Universe: Why Our Planet Is Unique* (Hoboken, NJ: Wiley, 2011). For a more technical discussion, see Anders Sandberg, Eric Drexler, and Toby Ord, "Dissolving the Fermi Paradox," arXiv preprint (June 6, 2018), arXiv:1806.02404.

For a review of exoplanet demographics, see B. Scott Gaudi, Michael Meyer, and Jessie Christiansen, "The Demographics of Exoplanets," 2020arXiv201104703G, in *ExoFrontiers: Big Questions in Exoplanetary Science*, edited by Nikku Madhusudhan (Bristol, UK: IOP Publishing, 2021).

On the critical role of water, see Darius Modirrousta-Galian and Giovanni Maddalena, "Of Aliens and Exoplanets: Why the Search for Life, Probably, Requires the Search for Water," *JBIS: Journal of the British Interplanetary Society* (2021), arXiv:2104.01683. See also H. K. Vedantham et al., "Coherent Radio Emission from a Quiescent Red Dwarf Indicative of Star-Planet Interaction," *Nature Astronomy* 4 (2020): 577–83, https://www.nature.com/articles/s41550-020-1011-9.

For an introduction to LUVOIR, see LUVOIR Team, "The LUVOIR Mission Concept Study Interim Report," eprint (September 2018), https://ui.adsabs.harvard .edu/abs/2018arXiv180909668T. See also The Optical Society, "New Telescope Design Could Capture Distant Celestial Objects with Unprecedented Detail" (press release), Astrobiology Web, March 19, 2020, http://astrobiology.com/2020/03/new -telescope-design-could-capture-distant-celestial-objects-with-unprecedented-detail .html.

For further reading, see the following:

Donovan Alexander, "Invisibility Cloaks Are No Longer Just Science Fiction," Interesting Engineering, February 19, 2021, https://interestingengineering.com /invisibility-cloaks-are-no-longer-just-science-fiction.

Lee Billings, "Alien Supercivilizations Absent from 100,000 Nearby Galaxies," *Scientific American*, April 17, 2015, https://www.scientificamerican.com/article/alien -supercivilizations-absent-from-100-000-nearby-galaxies/.

Nadia Drake, "Promising Sign of Life on Venus Might Not Exist After All," *National Geographic*, October 23, 2020, https://www.nationalgeographic.com /science/article/venus-might-not-have-much-phosphine-dampening-hopes -for-life.

Andreas M. Hein et al., "The Andromeda Study: A Femto-Spacecraft Mission to Alpha Centauri," August 11, 2017, https://arxiv.org/abs/1708.03556.

Elizabeth Kolbert, "Have We Already Been Visited by Aliens?," *The New Yorker*, January 18, 2021, https://www.newyorker.com/magazine/2021/01/25/have-we -already-been-visited-by-aliens.

Avi Loeb, *Extraterrestrial: The First Sign of Intelligent Life beyond Earth* (Boston: Houghton Mifflin Harcourt, 2021).

C. R. Nave, "The Moon's Role in the Habitability of the Earth," HyperPhysics, Georgia State University, https://ef.engr.utk.edu/hyperphysics/hbase/Astro/moonhab.html.

Luigi Secco, Marco Fecchio, and Francesco Marzari, "Habitability on Local, Galactic, and Cosmological Scales," December 2, 2019, arXiv:1912.01569.

V. I. Slysh, "A Search in the Infrared to Microwave for Astroengineering Activity," in *The Search for Extraterrestrial Life: Recent Developments*, proceedings of the symposium, Boston, June 18–21, 1984, A86-38126 17-88 (Dordrecht: D. Reidel Publishing Co., 1985), 315–19, 1985IAUS..112..315S.

Shannon Stirone, Kenneth Chang, and Dennis Overbye, "Life on Venus? Astronomers See Phosphine Signal in Its Clouds," *New York Times*, September 14, 2020, https://www.nytimes.com/2020/09/14/science/venus-life-clouds.html.

Chapter 8. Survival

Michael J. Benton, "The Origins of Modern Biodiversity on Land," *Philosophical Transactions of the Royal Society B: Biological Sciences* 365, no. 1558 (November 2010): 3667–79, https://doi.org/10.1098/rstb.2010.0269.

Nick Bostrom, *Superintelligence: Paths, Dangers, Strategies* (New York: Oxford University Press, 2014).

Deborah Byrd, "New Odds on Asteroid Bennu. Will It Strike Earth?," EarthSky, August 12, 2021, https://earthsky.org/space/new-odds-on-asteroid-bennu/.

Freeman John Dyson, "Search for Artificial Stellar Sources of Infrared Radiation," *Science* 131, no. 3414 (June 3, 1960): 1667–68.

John Ellis et al., "Review of the Safety of LHC collisions," *Journal of Physics G: Nuclear and Particle Physics* 35 (2008): 115004, arXiv:0806.3414.

Roger L. Griffith et al., "The Ĝ Infrared Search for Extraterrestrial Civilizations with Large Energy Supplies, III. The Reddest Extended Sources in WISE," *Astrophysical Journal Supplement Series* 217, no. 2 (April 2015): article id. 25, 2015ApJS..217 . . . 25G.

Joseph I. Kapusta, "Accelerator Disaster Scenarios, the Unabomber, and Scientific Risks," *Physics in Perspective* 10, no. 2 (2008): 163–81, arXiv:0804.4806.

N. S. Kardashev, "Transmission of Information by Extraterrestrial Civilizations," *Soviet Astronomy* 8 (October 1964): 217, https://ui.adsabs.harvard.edu/abs/1964SvA.....8..217K.

Ray Kurzweil, *The Singularity Is Near: When Humans Transcend Biology* (New York: Penguin, 2006).

John Menick, "Move 37: Artificial Intelligence, Randomness, and Creativity," *Mousse*, January 4, 2016, http://moussemagazine.it/john-menick-ai-1-2016/.

John Menick, "Move 37: Artificial Intelligence, Randomness, and Creativity: Part 2," *Mousse*, March 11, 2016, http://moussemagazine.it/john-menick-ai-2-2016/.

"A Mysterious Wasteland," *Time*, May 12, 1986, http://content.time.com/time/subscriber/article/0,33009,961334,00.html.

Toby Ord, *The Precipice: Existential Risk and the Future of Humanity* (New York: Hachette Books, 2020).

Steven J. Ostro and Carl Sagan, "Cosmic Collisions and the Longevity of Non-Spacefaring Galactic Civilizations," *Astrodynamics and Geophysics* (May 1998), http://hdl.handle.net/2014/19498.

Michael E. Peskin, "The End of the World at the Large Hadron Collider?," *Physics* 1, no. 14 (August 18, 2008), https://physics.aps.org/articles/v1/14.

Chris Phoenix and Eric Drexler, "Safe Exponential Manufacturing," *Nanotechnology* (Institute of Physics Publishing) 15 (2004): 869–72, http://crnano.org/IOP%20-%20Safe%20Exp%20Mfg.pdf.

David Silver et al., "Mastering Chess and Shogi by Self-Play with a General Reinforcement Learning Algorithm," December 5, 2017, arXiv:1712.01815.

Hector Socas-Navarro et al., "Concepts for Future Missions to Search for Techno-signatures," *Acta Astronautica* 182 (May 2021): 446–53, 2021AcAau.182..446S.

Diane Soran and Danny Stillman, "An Analysis of the Alleged Kishtym Disaster," Los Alamos National Lab report LA-9217-MS, U.S. Department of Energy, Office of Scientific and Technical Information, January 1, 1982, https://www.osti.gov/servlets/purl/5254763-UCvDE3/.

Alexey Turchin and David Denkenberger, "Classification of Global Catastrophic Risks Connected with Artificial Intelligence," *AI & Society* 35 (2020): 147–63.

Alex Wellerstein, "Manhattan Project," *Encyclopedia of the History of Science*, Carnegie Mellon University, April 2019, https://doi.org/10.34758/swph-yq79.

Jason T. Wright, "Strategies and Advice for the Search for Extraterrestrial Intelligence," *Acta Astronautica* 188 (November 2021): 203–14, arXiv:2107.07283.

Jason T. Wright and Steinn Sigurdsson, "Families of Plausible Solutions to the Puzzle of Boyajian's Star," *Astrophysical Journal Letters* 829, no. 1 (2016): article id. L3, 2016ApJ ... 829L ... 3W.

Chapter 9. Internationalization

Bradley C. Edwards, "The Space Elevator," 2000, http://www.niac.usra.edu/files/library/meetings/annual/jun02/521Edwards.pdf.

Martin Elvis, "What Can Space Resources Do for Astronomy and Planetary Science?," *Space Policy* 37, pt. 2 (August 2016): 65–76, https://www.sciencedirect.com/science/article/abs/pii/S0265964616300406.

Elvis, Martin, Alanna Krolikowski, and Tony Milligan, "Concentrated Lunar Resources: Implications for Governance and Justice," *Philosophical Transactions of the Royal Society A* 379, no. 2188 (January 2021): article id. 20190563, https://ui.adsabs.harvard.edu/abs/2021RSPTA.37990563E.

Thomas Marshall Eubanks and Charles F. Radley, "Scientific Return of a Lunar Elevator," *Space Policy* 37, pt. 2 (August 2016): 97–102, https://doi.org/10.1016/j.spacepol.2016.08.005.

Namrata Goswami, "China's Grand Strategy in Outer Space: To Establish Compelling Standards of Behavior," *The Space Review*, August 5, 2019, https://www.thespacereview.com/article/3773/1.

George F. Sowers, "A Cislunar Transportation System Fueled by Lunar Resources," *Space Policy* 37, pt. 2 (August 2016): 103–9, https://www.sciencedirect.com/science/article/abs/pii/S0265964616300352.

United Nations Office for Outer Space Treaty, "Treaty on Principles Governing the Activities of States in the Exploration and Use of Outer Space, Including the Moon and Other Celestial Bodies," 1966, http://www.unoosa.org/oosa/en/ourwork/spacelaw/treaties/introouterspacetreaty.html.

Chapter 10. The Next Century

The quote by Stephen Hawking comes from "Surprising Science: #5: Stephen Hawking's Warning: Abandon Earth—Or Face Extinction," *Big Think*, July 27, 2010, https://bigthink.com/surprising-science/5-stephen-hawkings-warning-abandon-earth-or-face-extinction/.

I review the prospects for doing fundamental cosmology on the Moon in Joseph Silk, "The Limits of Cosmology: Role of the Moon," in "Astronomy from the Moon: The Next Decades" (special issue), *Philosophical Transactions of the Royal Society A* 379, no. 2188 (January 11, 2021), https://doi.org/10.1098/rsta.2019.0561.

For further reading, see the following:

Ian A. Crawford, "The Long-Term Scientific Benefits of a Space Economy," *Space Policy* 37, pt. 2 (August 2016): 58–61, https://www.sciencedirect.com/science/article/abs/pii/S0265964616300327.

Philip T. Metzger, "Space Development and Space Science Together, an Historic Opportunity," *Space Policy* 37, pt. 2 (August 2016): 77–91, https://www.sciencedirect.com/science/article/abs/pii/S0265964616300625.

Antonio Paris, "Physiological and Psychological Aspects of Sending Humans to Mars," *Journal of the Washington Academy of Sciences* 100, no. 4 (Winter 2014): 3–20, https://www.jstor.org/stable/jwashacadscie.100.4.0003?seq=1.

INDEX